THE ROLE OF BIOTECHNOLOGY IN EXPLORING AND PROTECTING AGRICULTURAL GENETIC RESOURCES

THE ROLE OF BIOTECHNOLOGY IN EXPLORING AND PROTECTING AGRICULTURAL GENETIC RESOURCES

Edited by

John Ruane

&

Andrea Sonnino

DISCOVERY PUBLISHING HOUSE

NEW DELHI-110002

"Published by arrangement with the Food and Agriculture Organization of the United Nations by Discovery Publishing House FOR SALE AND DISTRIBUTION IN INDIA ONLY

"This Book was Originally Published by the Food and Agriculture Organization of the United Nations (FAO).

First Indian Edition–2007

ISBN 978-81-8356-190-7

Published by

DISCOVERY PUBLISHING HOUSE
4831/24, Ansari Road, Prahlad Street, Darya Ganj, New Delhi-110002 (India)
Phone: 23279245 • Fax: 91-11-23253475
E-mail: dphbooks@rediffmail.com, dphtemp@indiatimes.com

Printed at:
Sachin Printers, Delhi

Contents

Contributors

Gianni Barcaccia
Dipartimento di Agronomia Ambientale e Produzioni Vegetali (DAAPV)
Plant Genetics and Genomics
Via dell'Università 16
35020 Legnaro (Padova), Italy
gianni.barcaccia@unipd.it

Devin M. Bartley
FAO Inland Water Resources and Aquaculture Service
Rome, Italy
devin.bartley@fao.org

Ricardo A. Cardellino
FAO Animal Production Service
Rome, Italy
ricardo.cardellino@fao.org

M. Carmen de Vicente
International Plant Genetic Resources Institute (IPGRI)
Office for the Americas
c/o CIAT, A.A. 6713
Cali, Colombia
c.devicente@cgiar.org

Jan Engels
International Plant Genetic Resources Institute (IPGRI)
Via dei Tre Denari, 472/a, 00057
Maccarese, Rome, Italy
j.engels@cgiar.org

Marcio Elias Ferreira
Catholic University of Brasilia (UCB)
EMBRAPA Genetic Resources and Biotechnology (CENARGEN)
Parque Estação Biológica 70770-900
Caixa Postal 02372 - Brasília, Brazil
ferreira@cenargen.embrapa.br

Brad Fraleigh
FAO Seed and Plant Genetic Resources Service
Rome, Italy
brad.fraleigh@fao.org

Felix Alberto Guzmán
International Plant Genetic Resources Institute (IPGRI)
Office for the Americas
c/o CIAT, A.A. 6713
Cali, Colombia
f.a.guzman@cgiar.org

Olivier Hanotte
International Livestock Research Institute (ILRI)
P.O. Box 30709
Nairobi 00100, Kenya
o.hanotte@cgiar.org

Sipke Joost Hiemstra
Centre for Genetic Resources, the Netherlands (CGN)
Wageningen University and Research Centre
PO Box 65
8200 AB Lelystad, Netherlands
sipkejoost.hiemstra@wur.nl

Han Jianlin
International Livestock Research Institute (ILRI)
P.O. Box 30709
Nairobi 00100, Kenya
h.jianlin@cgiar.org

Maurizio Lambardi
Istituto per la Valorizzazione del Legno e delle Specie Arboree (IVALSA)
National Research Council (CNR), Polo Scientifico
Via Madonna del Piano
50019 Sesto Fiorentino (Firenze), Italy
lambardi@ivalsa.cnr.it

Sergio Lanteri
DiVaPra , Plant Genetics and Breeding
Via L. da Vinci 44,
10095 Grugliasco (Torino), Italy
sergio.lanteri@unito.it

Michele Morgante
Dipartimento di Scienze Agrarie ed Ambientali
Via delle Scienze 208
33100 Udine, Italy
michele.morgante@uniud.it

Bart Panis
Laboratory for Tropical Crop Improvement, K.U. Leuven
Kasteelpark Arenberg 13
3001 Leuven, Belgium
bart.panis@agr.kuleuven.ac.be

Ajay Parida
M.S. Swaminathan Research Foundation
Third Cross Road, Taramani Institutional Area
Chennai-600113, India
ajay@mssrf.res.in

Craig Primmer
Division of Genetics and Animal Physiology
Department of Biology, 20014
University of Turku, Finland
craig.primmer@utu.fi

Prashanth S. Raghavan
M.S. Swaminathan Research Foundation
Third Cross Road, Taramani Institutional Area
Chennai-600113, India
sequencing@mssrf.res.in

V. Ramanatha Rao
International Plant Genetic Resources Institute (IPGRI)
Office for Asia, the Pacific and Oceania
P.O. Box 236, UPM Post Office, Serdang
43400 Selangor Darul Ehsan, Malaysia
v.rao@cgiar.org

Jonathan Robinson
FAO Consultant
FAO Working Group on Biotechnology
Rome, Italy
jrobinson@tiscalinet.it

John Ruane
FAO Working Group on Biotechnology
Rome, Italy
john.ruane@fao.org

Pierre Sigaud
FAO Forest Resources Development Service
Rome, Italy
pierre.sigaud@fao.org

Henner Simianer
Institute of Animal Breeding and Genetics
Georg-August-University
Goettingen, Germany
hsimian@gwdg.de

Andrea Sonnino
FAO Research and Technology Development Service
Rome, Italy
andrea.sonnino@fao.org

Tette van der Lende
Animal Sciences Group
Wageningen University and Research Centre
PO Box 65
8200 AB Lelystad, Netherlands
tette.vanderlende@wur.nl

Giovanni G. Vendramin
Istituto di Genetica Vegetale
National Research Council (CNR)
Via Madonna del Piano
50019 Sesto Fiorentino (Firenze), Italy
giovanni.vendramin@igv.cnr.it

Henri Woelders
Centre for Genetic Resources (CGN) and Animal Sciences Group of
Wageningen University and Research Centre
PO Box 65
8200 AB Lelystad, Netherlands
henri.woelders@wur.nl

Preface

One of FAO's major roles is to provide member countries and their institutions with factual, comprehensive and current information on issues related to food and agriculture, including biotechnology applications. It provides this information through the FAO Web site on biotechnology, an e-mail newsletter *FAO-BiotechNews*, and a series of e-mail conferences hosted by the FAO Biotechnology Forum. This e-mail-based forum was launched in 2000 with the aim of providing quality, balanced information on agricultural biotechnology in developing countries and to make a neutral platform available for people to exchange views and experiences on this subject. It has around 3 000 members worldwide, and in its first five years it has hosted 13 moderated e-mail conferences.

Each conference of the FAO Biotechnology Forum takes one particular topic and discusses it for a limited amount of time, normally four weeks. Before each conference begins, a background document is sent to the Forum members, which gives a good background to the conference theme, in a balanced neutral way, and is written in easily understandable language so that people with little knowledge of the area may understand what the theme is about. The conferences are moderated, open to everyone, and normally 350 to 650 people join in. Despite the fact that there are tremendous global inequalities in use of the Internet, there has been very active participation from developing countries, with roughly 50 percent of all messages posted coming from people living in these countries. After each conference, a summary document is sent to the Forum members, which summarizes the main issues discussed during the e-mail conference, based on the messages posted by the participants during the conference. Although it is an e-mail-based forum, all documents and messages are also made available on the Internet (www.fao.org/biotech).

On the occasion of World Food Day 2004, the United Nations Secretary-General Kofi Annan urged "individuals and institutions alike to give greater attention to biodiversity as a key theme in our efforts to fight the twin scourges of hunger and poverty and achieve the Millennium Development Goals". He also noted that the unprecedented loss of biodiversity over the past century was a major cause for alarm, where

> "many freshwater fish species, which can provide crucial dietary diversity to the poorest households, have become extinct, and many of the world's most important marine fisheries have been decimated. Food supplies have also been made more vulnerable by our reliance on a very small number of species: just 30 crop species dominate food production and 90 per cent of our animal food supply comes from just 14 mammal and bird species - species which themselves rely on biodiversity for their productivity and survival. There has been a substantial reduction in crop

genetic diversity in the field and many livestock breeds are threatened with extinction".

On the same occasion, FAO's Director-General Jacques Diouf also underlined that although forests are among the world's most important repositories of biological diversity, the world forest cover is decreasing at an alarming rate.

It is in this context of declining agricultural biodiversity that the FAO Biotechnology Forum decided to dedicate Conference 13 to "the role of biotechnology for the characterization and conservation of crop, forest, animal and fishery genetic resources in developing countries", which took place between 6 June and 4 July 2005. Biotechnology is a broad collection of tools, which can be applied for a range of different purposes (e.g. genetic improvement of populations; disease diagnosis and vaccine development; improvement of feeds). The focus in the e-mail conference was on biotechnology tools, such as molecular markers or cryopreservation and reproductive technologies that can be used directly for the characterization and/or conservation of genetic resources for food and agriculture. About 650 people subscribed to the conference and 127 messages were posted from people living in 38 different countries. Over 60 percent of messages posted were from people living in developing countries. As part of the preparations for the conference, an international workshop was held from 5 to 7 March 2005 in Turin, Italy, on the same subject that was organized by the FAO Working Group on Biotechnology in collaboration with the Fondazione per le Biotecnologie, the Econogene project (a European Union-funded project on biodiversity and conservation of sheep and goat breeds in marginal areas) and the Italian Society of Agriculture Genetics.

In this book, we are happy to bring together papers from the meeting in Turin and the background and summary documents from the subsequent e-mail conference. The book aims to provide an updated overview of the current status of the world's genetic resources for food and agriculture, of the use of biotechnology tools for characterizing and conserving these genetic resources, and of the many specific issues involved in applying them in developing countries. Section I contains four papers on the status of the world's livestock, fishery, crop and forest genetic resources. Section II considers the use of cryopreservation and reproductive technologies for conservation of animal and plant genetic resources. Section III is dedicated to the use of molecular markers for characterization and conservation of genetic resources. Finally, Section IV contains the two documents from the e-mail conference. We hope the book will be useful to individuals interested in the utilization of biotechnology for characterization and conservation of genetic resources for the good of humanity today and tomorrow.

Shivaji Pandey

Chairperson, FAO Working Group on Biotechnology

Acknowledgements

Chapters 1 to 14 of this book are based on papers presented at Sessions I, II and IV of an international workshop held from 5 to 7 March 2005 entitled, *The Role of Biotechnology for the Characterisation and Conservation of Crop, Forestry, Animal and Fishery Genetic Resources*, organized by the FAO Working Group on Biotechnology (FAO-WGB), the Fondazione per le Biotecnologie and the Italian Society of Agriculture Genetics (SIGA). The workshop took place at the Villa Gualino Congress Center in Turin, Italy and would not have been the success it was without the efficient and scrupulous attention to detail of Claudia Mondino and Elena Spoldi and their co-workers from the Fondazione. The programme for the workshop was developed by Paolo Ajmone Marsan (Catholic University of Piacenza), Sergio Lanteri (University of Turin), Luigi Monti (University of Naples), Andrea Sonnino (FAO-WGB) and Fabio Veronesi (University of Perugia), with the support of John Ruane and helpful inputs from Devin Bartley, Kwame Boa-Amponsem, Elcio Guimarães, Irene Hoffman, Beate Scherf and Pierre Sigaud (all from the FAO-WGB). The support of SIGA, in particular of its President Michele Stanca, is also gratefully acknowledged in this context.

The remaining two chapters, 15 and 16, are from the e-mail conference organized by the FAO-WGB roughly three months after the Turin workshop. Comments from Christel Palmberg-Lerche and Pierre Sigaud on a previous draft of the background document to the conference (Chapter 15) are gratefully acknowledged. During the conference, about 650 people subscribed and 127 e-mail messages were posted from 64 people, whose comments and opinions formed the basis for Chapter 16. We are extremely grateful to all of the individuals who submitted messages, for devoting their time and effort to sharing their views, insights and experiences with the other Forum members. Barbara Hall is thanked for assistance with editorial and layout aspects. Finally, we wish to express special thanks to Jim Dargie, the former Chairperson of the FAO-WGB, and to his successor, Shivaji Pandey, for supporting these activities regarding the role of biotechnology tools for the characterization and conservation of genetic resources for food and agriculture.

Acronyms and abbreviations

ABA	Abscisic acid
AB-QTL	Advanced back cross quantitative trait locus
AFLPs	Amplified fragment length polymorphism markers
AI	Artificial insemination
AnGR	Animal genetic resources
BAC	Bacterial artificial chromosones
BECA	Biosciences Eastern and Central Africa
CBD	Convention on Biological Diversity
cDNA	Complementary DNA
CGIAR	Consultative Group on International Agricultural Research
CGRFA	Commission on Genetic Resources for Food and Agriculture
CPA	Cryoprotective agents
CV	Coefficient of variation
EG	Ethylene glycol
EST	Expressed sequence tag
ESU	Evolutionary significant unit
FAO	Food and Agriculture Organization of the United Nations
FAO-WGB	FAO Working Group on Biotechnology
FGR	Forest genetic resources
GIFT	Genetic improvement of farmed tilapia
He	Expected heterozygosity
Ho	Observed heterozygosity
ICRAF	World Agroforestry Centre
ILRI	International Livestock Research Institute
INIBAP	International Network for Improvement of Banana and Plantain
IPGRI	International Plant Genetic Resources Institute
ISAG	International Society for Animal Genetics
ISSR	Inter-single sequence repeat
IUCN	World Conservation Union
IUFRO	International Union of Forest Research Organizations
LD	Linkage disequilibrum
mtDNA	Mitochondrial DNA
Ne	Effective population size
OECD	Organisation for Economic Co-operation and Development
PCR	Polymerase chain reaction
PGRFA	Plant genetic resources for food and agriculture
PIC	Polymorphic index content

QTL	Quantitative trait loci
RAPD	Random amplified polymorphic DNA
RFLP	Restriction fragment length polymoprhism
SAMPL	Selective amplification of microsatellite polymorphic loci
SIGA	Italian Society of Agriculture Genetics
SNP	Single nucleotide polymorphism
SoW-AnGR	State of the World's Animal Genetic Resources
SoW-PGRFA	State of the World's Plant Genetic Resources for Food and Agriculture
SSR	Simple sequence repeats
UNEP	United Nations Environment Programme
UNFF	United Nations Forum on Forests
UPGMA	Unweighted pair-group method arithmetic average

I. Status of the world's genetic resources for food and agriculture

1. Status of the world's livestock genetic resources: preparation of the first *Report on the State of the World's Animal Genetic Resources*

Ricardo A. Cardellino

1.1 SUMMARY

FAO's Commission on Genetic Resources for Food and Agriculture (CGRFA) is the intergovernmental forum for negotiating action on genetic resources for food and agriculture. This commission deals with all aspects of genetic resources including plant, animal, forest and fisheries, and holds regular sessions every two years. Technical guidance on aspects dealing with domestic animals is provided by a working group of this commission, the Intergovernmental Technical Working Group on Animal Genetic Resources. Member countries represented in these intergovernmental bodies have requested FAO to develop and implement a *global strategy for the management of farm animal genetic resources* as a strategic framework to guide international efforts in the animal genetic resources sector. Aims of the *global strategy* are to enhance awareness of the multiple roles and values of animal genetic resources, provide guidance for establishing national, regional and global policies, strategies and actions, and facilitate and coordinate the activities of many independent organizations that have an interest in animal genetic resources.

In the context of the *global strategy*, member countries have requested FAO to prepare the first *Report on the State of the World's Animal Genetic Resources*. Status on the progress of this process is reported in the present chapter.

1.2 FARM ANIMAL GENETIC RESOURCES AT RISK

Farm animal genetic resources face a double challenge. On the one hand, the demand for animal products is increasing in developing countries: FAO has estimated that demand for meat will double by 2030 (with respect to 2000); over the same thirty-year period demand for milk will more than double. On the other hand, animal genetic resources are disappearing rapidly worldwide. Over the past 15 years, 300 out of 6 000 breeds identified by FAO have become extinct. Many breeds of local importance for food security are not being improved or utilized in a sustainable manner and are in danger of being lost or diluted by crossbreeding.

Conservation and development of local breeds is important because many of them utilize lower quality feed, are more resilient to climatic stress and to local parasites and diseases, and represent a unique source of genes for improving health and performance traits of industrial breeds. It is also important to develop and utilize local breeds that are already adapted to their environments, most of which are harsh, with very limited natural and managerial input. Animals genetically adapted to these conditions are expected to be more productive at lower costs, support food, agriculture and cultural diversity, and be effective in achieving local food security objectives.

Local communities depend on these adapted genetic resources in many countries. Their disappearance or drastic modification, for example, by crossbreeding, absorption or replacement by exotic breeds, will have serious negative impacts on these human populations. Presently, most breeds at risk of extinction are not supported by any established conservation programmes or active conservation through sustainable utilization (breeding plans) and therefore breed extinction rates are increasing globally.

1.3 THE GLOBAL STRATEGY FOR THE MANAGEMENT OF FARM ANIMAL GENETIC RESOURCES

The key component of the *global strategy* is the country-based planning and implementation infrastructure, which includes five structural elements:

- The *global focal point* at FAO headquarters leads the planning, development and implementation of the overall strategy; develops and maintains the information and communication systems; oversees preparation of guidelines; coordinates regional activity; prepares reports and documents for meetings; facilitates policy discussions; identifies training, education and technology transfer needs; develops programme and project proposals; and mobilizes donor resources.
- *Regional focal points* facilitate regional communications; provide technical assistance and leadership; coordinate regional training, research and planning activities; help develop regional policies; assist in identifying project priorities and proposals, and interact with government agencies, donors, research institutions and non-governmental organizations.
- *National focal points* lead, facilitate and coordinate country activities; identify capacity-building needs; develop project proposals; assist with the development and implementation of country policies; and interface with national stakeholders, regional focal points and the global focal point.
- *Donor and stakeholder involvement* is necessary to provide financial and institutional support to the *global strategy*. In this context, the global focal point seeks to ensure stakeholder involvement in all major aspects of the *global strategy*, facilitating opportunities for governmental and non-governmental contributions.
- *DAD-IS, the Domestic Animal Diversity Information System* (www.fao.org/dad-is), is a widely available and easily accessible global database and information source. This global facility makes it possible to share data and information among countries, allowing a rapid and cost-effective distribution of

guidelines, reports and meeting documents, and provides a platform to exchange views and address specific information requests, linking breeders, scientists and policy-makers. A key feature of DAD-IS is the breeds database, which provides the data for the early warning system for animal genetic resources through the World Watch List for Domestic Animal Diversity, whose third edition was released in 2001.

1.4 FIRST *REPORT ON THE STATE OF THE WORLD'S ANIMAL GENETIC RESOURCES*

As part of the *global strategy for the management of farm animal genetic resources,* FAO invited 188 countries to participate in the first *Report on the State of the World's Animal Genetic Resources*, which is to be completed by 2006. To date, 151 countries have accepted to submit country reports. Guidelines for preparation of country reports have been published in Animal Genetic Resources Information Bulletin (FAO) no. 30. These guidelines are used to assist countries in preparing reports as strategic policy documentation covering the state of animal genetic resources, the state of the art and national capacity to manage these resources, and country needs and priorities. Country reports will serve as the base documentation for the State of the World Reporting Process; thus the involvement of all stakeholders in the development of these reports is strongly encouraged.

The objective of the country and global assessments is to provide a comprehensive analysis of the status and trends of the world's farm animal biodiversity and of their underlying causes, as well as of local knowledge regarding its management. The task is to go beyond description of the resources by analysing the state of these resources and the capacities to manage them, drawing lessons from past experiences and identifying problems and priorities. Country reports are policy documents covering three strategic questions: *Where are we? Where do we need to be? How do we get to where we need to be?* Country reports are intended to be used in planning and implementing priority country actions. In addition, the country report will serve as documentation for the development of the regional and global reports on strategic priorities for action and, subsequently, the first *Global Report on the State of Farm Animal Genetic Resources.*

Country reports provide an assessment in three major areas:

- the *state of diversity* to evaluate the state of conservation, erosion and utilization of farm animal agricultural biodiversity, and an analysis of the underlying processes;
- the *state of national capacity* to manage animal genetic resources, including existing policies, management plans, institutional infrastructures, human resources and equipment;
- the *state of the art* of the available methodologies and technologies to assist farmers, breeders and scientists to better understand, use, develop and conserve animal genetic resources and thereby contribute to global food security and rural development.

International organizations are also being invited to contribute to the state of the world's animal genetic resources preparatory process by providing reports. The long-term aim of the process is for countries and regions to build on the analyses contained in the country reports in order to plan and implement appropriate management of their farm animal genetic resources.

The first *Report on the State of the World's Animal Genetic Resources* will contain the *Report on Strategic Priorities for Action* and will be based on a synthesis of country reports, thematic studies and reports from international organizations.

1.5 FIELDWORK AT COUNTRY AND REGIONAL LEVELS

Countries were requested to nominate a national focal point and designate a national coordinator to facilitate the development of the country network on the management of animal genetic resources and to serve as official contact with the global focal point. Keeping in mind that the process involves both scientific and policy matters, the establishment of a National Consultative Committee is recommended to identify the primary areas and issues that need to be addressed in the preparation of the country report and to oversee its preparation. It is essential that the National Consultative Committee have wide and diverse representation and develop a broad network to ensure opportunities for all stakeholders to contribute to the country report.

The response of countries to the invitation of FAO's Director-General to participate in the first *Report on the State of the World's Animal Genetic Resources* and submit a country report has been very positive. During part of 2001 and 2002, FAO trained almost 400 professionals from 178 countries in the preparation of national reports. At the moment, FAO has a team of 15 consultants working in 14 country groupings in all regions of the world. Most countries have undertaken the organization of national stakeholder workshops to elaborate their animal genetic resources policies leading to the country reports. Table 1 shows the regional distribution of country reports submitted to FAO.

FAO has organized 14 subregional workshops to discuss draft country reports and regional priorities for action. This has promoted regional cooperation and allows countries that may be experiencing delays to catch up with those in a more advanced state of country report preparation and learn from their experiences. These sessions were coordinated by the regional facilitators acting as FAO consultants.

FAO has provided technical and financial support to 115 countries with contributions from the Governments of the Netherlands and Finland and from the Nordic Gene Bank. FAO and the World Association for Animal Production (WAAP) signed an agreement to provide technical and operational support for the state of the world animal genetic resources reporting process, including training and country follow-up. FAO considers this cooperation a prime example of effective collaboration with an international non-governmental organization.

TABLE 1
Numbers of draft and final country reports already submitted and to be submitted to FAO (15 February 2005)

No. of country reports	Africa	Asia-Pac.	Europe	LatAmCar	Near East	N. America	Non-FAO	Total
Drafts	6	5	3	6	3	0	1	24
Finals	38	26	35	16	10	2	2	129
To be submitted	3	2	2	10	0	0	0	17
Total no. of reports	47	33	40	32	13	2	3	170
Total no. of countries	48	39	44	33	21	2	3	190
Expected result (%)	97.92	84.62	90.91	96.97	61.90	100.00	100.00	

1.6 FINALIZATION OF THE FIRST REPORT ON THE STATE OF THE WORLD'S ANIMAL GENETIC RESOURCES

2005 Using as a basis for discussion the first draft of the *Report on Strategic Priorities for Action*, FAO will convene regional consultations to review and determine regional priorities, identify funding options and expose gaps where international assistance is required. Such consultations will depend to a large extent on the availability of extra-budgetary resources.

FAO will prepare a draft of the first *Report on the State of the World's Animal Genetic Resources*. The results of the regional consultations, the available country reports, reports from international organizations and thematic studies will provide the basis for preparing the first draft of the first report by the end of 2005.

2006 A review of the first draft of the first *Report on the State of the World's Animal Genetic Resources* will be undertaken by governments and stakeholders in the first half of 2006. A second Global Workshop for National Coordinators and a stakeholders' meeting will be convened to undertake a comprehensive technical review of the draft early in 2006. The Intergovernmental Technical Working Group on Animal Genetic Resources will meet in 2006 in order to review the first draft of the first *Report on the State of the World's Animal Genetic Resources*, evaluate the operation of the follow-up mechanism and prepare a draft agenda for an *Intergovernmental Technical Conference on Animal Genetic Resources* to be held in 2007. The CGRFA, at its Eleventh Regular Session in 2006, will review the first draft of the first *Report on the State of the World's Animal Genetic Resources*, evaluate the follow-up mechanism, and endorse an agenda for the first *Intergovernmental Technical Conference on Animal Genetic Resources.*

1.7 RECOMMENDATIONS OF THE COMMISSION ON GENETIC RESOURCES FOR FOOD AND AGRICULTURE

The Commission made a series of recommendations to FAO (Table 2) based on three main elements:

- the completion of the state of the world's animal genetic resources process;
- the establishment of a follow-up mechanism with the following objectives:
  - mobilizing financial resources;
  - providing support in project design, development and submission to relevant funding agencies;
  - raising global awareness of the roles and values of animal genetic resources and their contribution to food and agriculture;
- the continued development of *the global strategy for the management of farm animal genetic resources.*

1.8 SPECIAL STUDIES

The following thematic studies are being conducted for inclusion in the *Report on the State of the World's Animal Genetic Resources*:

- valuation of animal genetic resources;
- community-based management of local animal genetic resources;

TABLE 2
Summary of recommendations of the Tenth Regular Session of the Commission on Genetic Resources for Food and Agriculture (8-12 November 2004)

Task	Overall activity
Assist countries at local and national levels, strengthen national focal points, implement concrete actions in countries, involve policy-makers.	Regional networking
Establish and promote sustainable regional focal points, support the informal network of regional facilitators.	Regional networking
Facilitate regional training in conservation and sustainable utilization of animal genetic resources.	Regional networking
Conduct regional consultations to discuss and endorse regional priorities for action.	Regional and national policy level
Establish a follow-up mechanism with a national and a regional focus.	Planning the follow-up mechanism
Develop and present detailed operational plan for the Report on the State of the World's Animal Genetic Resources.	Planning Report on the State of the World's Animal Genetic Resources
Writing the Report on the State of the World's Animal Genetic Resources as a platform for policy discussion and public awareness.	Draft of the Report on the State of the World's Animal Genetic Resources
Further develop the Report on Strategic Priorities for Action.	Draft of the Report on Strategic Priorities for Action
Develop decision-support tools for breeding programmes.	Animal breeding plans
Develop DAD-IS.	Data base and information
Prepare a proposal for monitoring system.	Conceptual development
Develop conceptual approach to conservation.	International seminars
Develop a plan for the International Technical Conference to be held in 2007.	Planning for 2007
Seek funding for follow-up mechanism.	Fundraising
Seek funding for the International Technical Conference.	Fundraising

- gene flow of major domestic animal species among countries and regions;
- options for the conservation of threatened farm animal populations;
- legal issues in management of animal genetic resources;
- biotechnology and animal genetic resources;
- measurement of domestic animal diversity (MoDAD);
- animal genetic resources and environment;
- impact of emergencies and interventions on animal genetic resources.

1.9 REGIONAL PRIORITIES IDENTIFIED BY COUNTRIES

The following is a list of regional priorities identified in regional workshops:

- institutional development and capacity building for the management of animal genetic resources;
- characterization and valuation of animal genetic resources;
- sustainable use and improvement of animal genetic resources;
- mainstreaming of animal genetic resources into national policies;
- development of national and multilateral legislation regarding animal genetic resources;
- conservation programmes, both *in situ* (breeding schemes) and *ex situ* (cryo-conservation);
- use of traditional knowledge in the management of animal genetic resources;
- cooperation in research and biotechnology aiming at reducing gaps between developing and developed countries;
- monitoring of animal genetic resources and developing emergency response mechanisms;
- enhancement of public awareness of the value and utilization of animal genetic resources.

1.10 REFERENCES

All cited bibliography and related publications can be found in www.fao.org/DAD-IS. All documentation pertaining to the Commission on Genetic Resources for Food and Agriculture (CGRFA) and the Intergovernmental Technical Working Group on Animal Genetic Resources can be found in www.fao.org/ag/cgrfa.

2. Status of the world's fishery genetic resources

Devin M. Bartley

2.1 SUMMARY

The world's fisheries are composed of 974 taxa of fin-fish, 143 crustaceans, 114 molluscs, 26 plants and 73 taxa of miscellaneous animals, while aquaculture production is composed of 153 species of fish, 60 molluscs, 44 crustaceans, 11 plants and several other miscellaneous taxa. These figures are certainly underestimates. Stocks of wild populations and farmed species exist, but are poorly documented for all but a few species.

2.2 WHAT ARE FISHERY GENETIC RESOURCES? WHAT NEEDS TO BE CHARACTERIZED?

The Convention on Biological Diversity (CBD) (1992) defines aquatic genetic resources as "(aquatic) genetic material of actual or political value" and genetic material as "any material of plant, animal, microbial or other origin containing functional units of heredity." Bartley and Pullin (1999) then asked the questions, "Are aquatic genetic resources, then, the sum total of all the aquatic plants, animals and micro-organisms on the planet? Does everything aquatic and alive and all of its DNA have actual or potential value?" This might indeed be the best assumption today because of the large knowledge gaps that remain in: (i) understanding how aquatic ecosystems function to support fisheries; (ii) how to choose aquatic species, for domestication; (iii) how to make rapid progress in domesticating them; and (iv) how to harness aquatic biochemicals and biological processes for the benefit of humankind. For the time being such a broad definition would make the terms "aquatic genetic resources" and "aquatic biodiversity" nearly synonymous. Therefore, there are different levels of genetic diversity, including ecosystems, communities, populations, genotypes and individual genes. Each level of the hierarchy has specific functions and supports the level above it (Bartley and Pullin, 1999). Since genetic resources have value in terms of economic, ecological and social uses, they need to be characterized. This is central to FAO's mandate concerning information, but is also vital for fishery management and aquaculture development. Both wild and farmed groups of fish need to be characterized. Aquaculture is the fastest growing food producing sector; by 2025 it is expected that one out of every two fish eaten will be from aquaculture. Although capture fisheries is the last major source of food derived from "hunting", fishing is important as an economic, social and cultural activity in much of the world.

This chapter reports on the status of aquatic species that are fished or farmed throughout the world. The number of species and production were derived from

FAO's databases. Unlike the terrestrial plant and animal sectors, there is no systematic effort to describe the state of the world's fishery genetic resources below the species level. Therefore, selected examples of well-studied groups or organisms are included to demonstrate the diversity and usefulness of fishery genetic resources.

2.3 CAPTURE FISHERIES

In 2002 FAO Members reported that 974 taxa of fin-fish, 143 taxa of crustaceans, 114 taxa of molluscs, 26 taxa of plants and 73 taxa of miscellaneous animals such as sea urchins, sea cucumbers and marine mammals were taken from the world's capture fisheries (Figure 1).

The 15 most productive fisheries in terms of quantity are listed in Table 3 with their production figures. Although over 1 000 taxa are represented in this data set, around ten species make up about one-third of total production. Overall production from the world's capture fishery increased up to the late 1980s and has now reached what most fishery scientists think is a plateau, that is, not much more production can be expected (Figure 2).

This information officially reported to FAO is certainly an underestimate of the number of species and genetic diversity contributing the world's capture fisheries. The categories "nei" (Table 3) refer to organisms "not elsewhere included", that is, generally, catch not identified to a species or to a major group. The "nei" groups account for over 20 percent of the global catch. Unfortunately, the information reported to FAO is getting worse in terms of reporting by species because now more than before, production is reported as coming from species "nei" (FAO, 2002).

FIGURE 1
Taxa reported taken from the world's capture fisheries in 2003

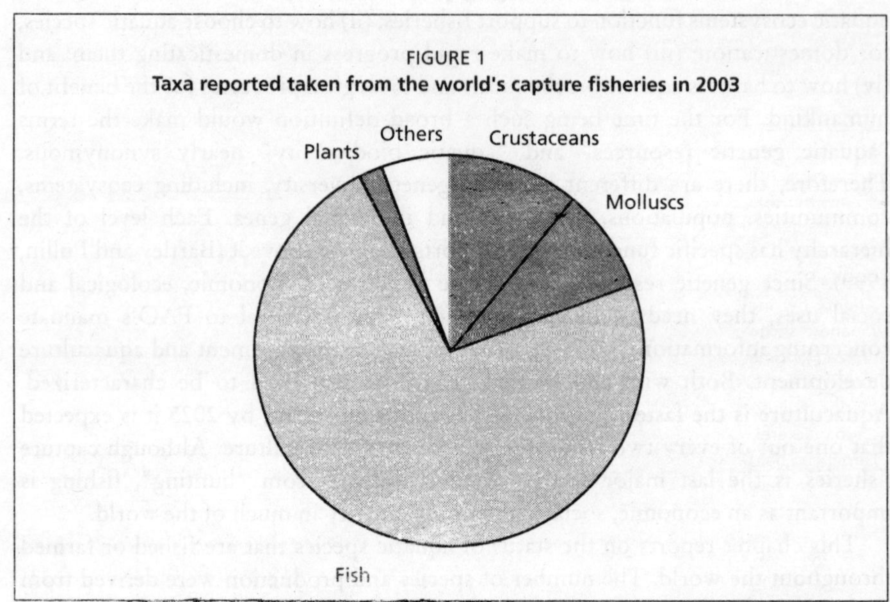

FIGURE 2
Production from the world's capture fisheries from 1950-2002

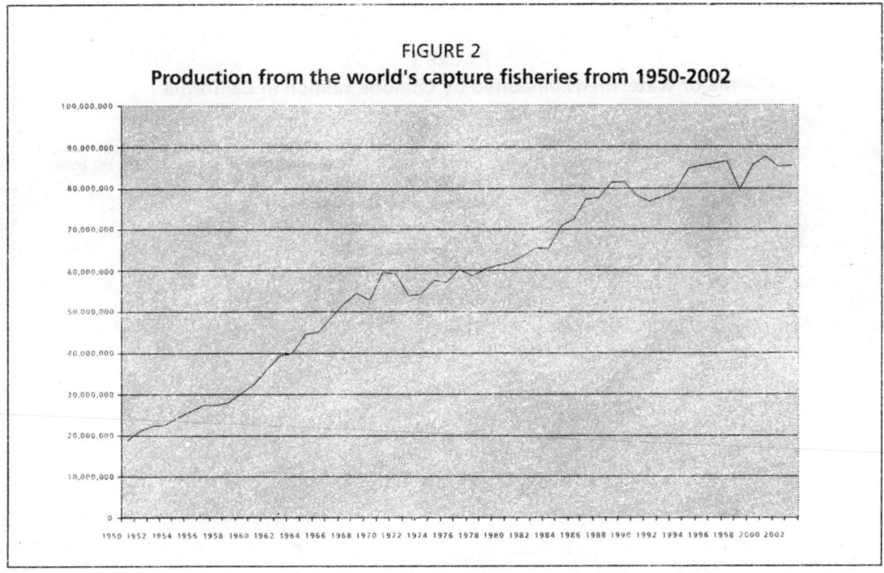

Below the species level, many of the world's fisheries are composed of numerous stocks. Definitions of what constitute a "stock" have not been agreed on. However, the term is used here to refer to a group of similar individuals within a species that preferentially breed within the group. Good examples of stocks are the various spawning migrations of salmon and trout. These runs or stocks can be differentiated spatially by river systems and temporally by seasons. For example, the chinook salmon fishery in the Pacific Northwest of North America is composed of hundreds

TABLE 3
The most important capture fisheries in terms of quantity in 2002

Taxa	Quantity (mt)	Percent	Cumulative
Marine fishes*	10,693,764	12.5	12.5
Anchoveta	9,702,614	10.1	23.8
Freshwater fishes*	4,389,297	5.1	29.0
Alaska pollock	2,654,854	3.1	32.1
Skipjack tuna	2,030,648	2.4	35.4
Capelin	1,961,724	2.3	36.7
Atlantic herring	1,872,013	2.2	38.9
Japanese anchovy	1,853,936	2.2	41.1
Chilean jack mackerel	1,750,078	2.0	43.1
Blue whiting	1,603,263	1.9	45.0
Marine molluscs*	1,491,849	1.7	46.8
Chub mackerel	1,470,673	1.7	48.4
Largehead hairtail	1,452,209	1.7	50.1
Marine crustaceans*	1,372,522	1.6	51.2
Yellowfin tuna	1,341,319	1.5	53.3

* nei = not elsewhere included; generally refers to catch not identified to species or group

Source: FAO FishStat Plus, 2005

FIGURE 3
Major watersheds inhabited by Chinook salmon in California

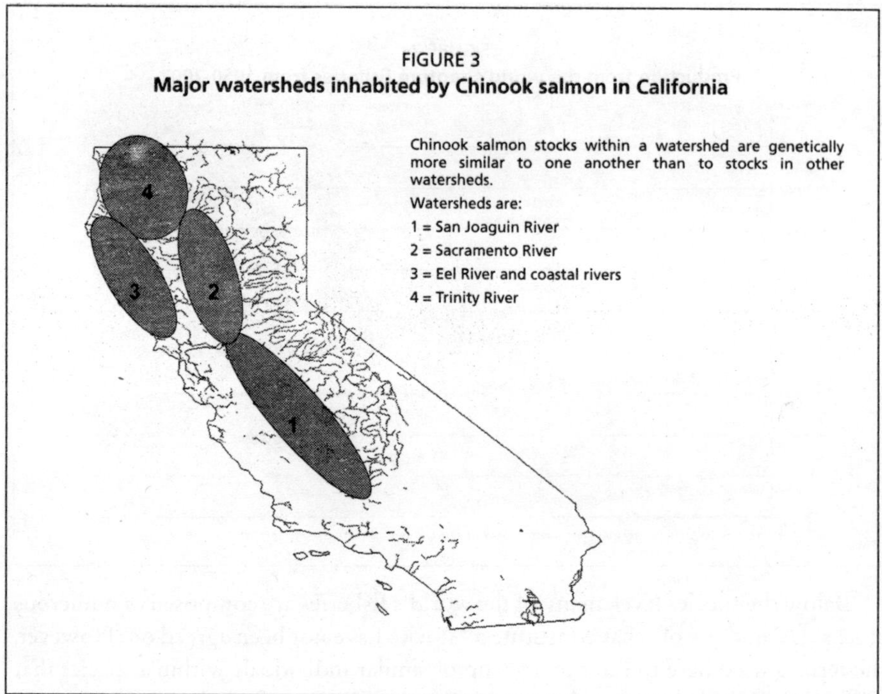

of genetically differentiated stocks that correspond to river systems and time of spawning (Figure 3) (Bartley *et al.*, 1992). The US Endangered Species Act has recognized the value of this intraspecific diversity and has afforded protection to numerous endangered runs under the Act. Although most of worlds' fisheries have not been genetically characterized to the extent that many salmonid fisheries have, stock structure has been found in many species. Genetic differentiation depends on a variety of factors and can be useful in setting management and conservation goals.

2.4 AQUACULTURE

In 2002 FAO Members reported that 153 species of fish, 60 species of molluscs, 44 species of crustaceans, 11 species of plants and several other miscellaneous taxa such as echinoderms, frogs, and crocodiles were farmed in various parts of the world (Figure 4). Contrary to the levelling in production from capture fisheries, aquaculture is expanding rapidly (Figure 5), especially in the developing world, and many governments have increased aquaculture development as development goals (Government of Kenya, 2003).

Aquaculture is a relatively new enterprise except for a few species, such as common carp that was domesticated several thousand years ago. Although there are a number of genetic techniques available to improve aquatic species (Table 4), the aquaculture sector lags behind the crop, livestock and poultry sectors in regard to the development of domesticated and genetically improved strains. However, progress is being made in domesticating species of fish such as rainbow and brown

FIGURE 4
Composition of the reported aquaculture production in 2002

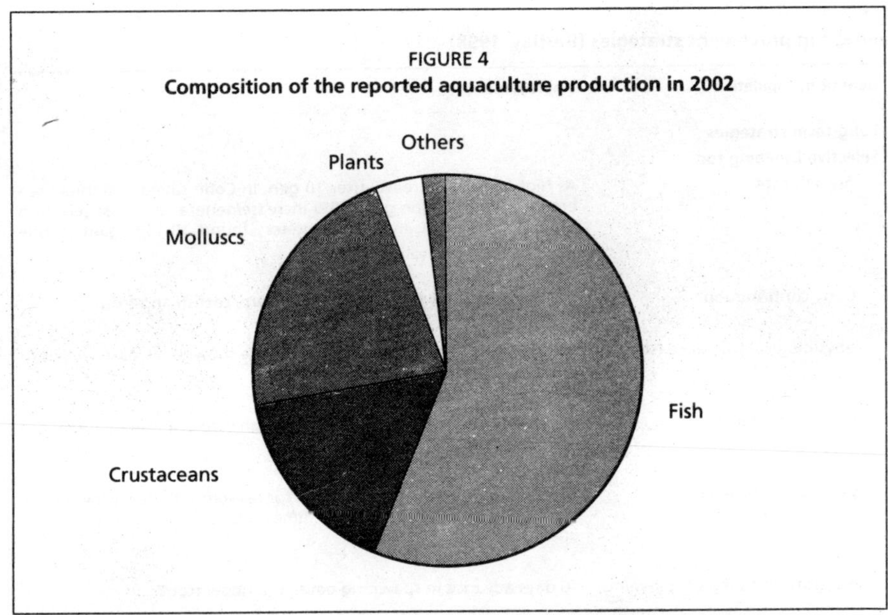

trout, Atlantic salmon, channel catfish, common carp, Chinese and Indian carps, and tilapia. While Pacific oyster and other oyster species have been genetically improved, very few crustaceans have been improved due to the problems in artificial breeding. Several projects have been developed or are currently in operation to characterize these important aquatic species.

For example, a European Union project SALMAP to characterize salmonids found 200 microsatellite markers for rainbow trout, 299 for Atlantic salmon and

FIGURE 5
Global aquaculture production, 1950-2002

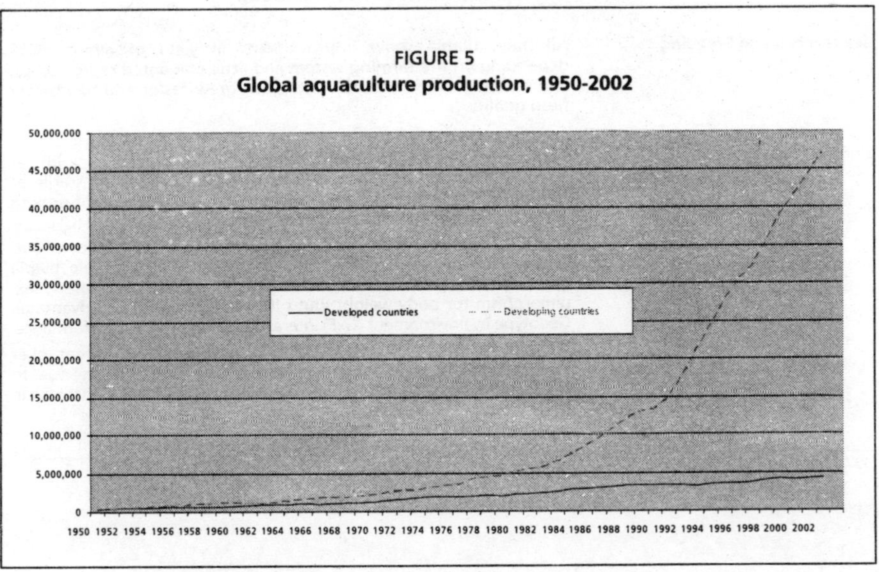

TABLE 4
Genetic improvement strategies (Bartley, 1998)

Genetic manipulation	Improvement
Long-term strategies	
Selective breeding for:	
growth rate	As high as 50% increase after 10 gen. in Coho salmon; gilthead sea bream mass selection gave 20% increase/generation; mass selection for live wt and SL in Chilean oysters found 10-13% gain in one generation.
body confirmation	High heritabilities found in common carp, catfish and trout.
physiological tolerance (stress)	Rainbow trout selected for high response showed increased levels of plasma cortisol levels.
disease resistance	Increased resistance to dropsy in common carp but disease resistance is difficult to select for.
pollutant resistance	Tilapia progeny from lines selected for resistance to heavy metals Hg, Cd, and Zn survived 3 to 5 times better than progeny from unexposed lines.
maturity and time of spawning	60 days advance in spawning date in rainbow trout
Gene transfer	Coho salmon with a growth hormone gene and promoter from Sockeye salmon grew 11 times (0-37 range) as fast as non-transgenics. Atlantic salmon grew 400% faster than normal during the first year.
Short-term strategies	
Intra-specific crossbreeding	Heterotic growth is seen in 55% and 22% of channel catfish and rainbow trout crosses, respectively. Chum salmon and largemouth bass showed no heterosis.
	Heterosis for wild x hatchery *S. aurata*; crossbreeds of channel catfish common carp showed 30-60% heterosis.
Sex reversal and breeding	All male tilapia show improvements in yield of almost 60% depending on the farming system and little unwanted reproduction and stunting. All female rainbow trout grew faster and had better flesh quality.
Chromosome manipulation	Pagrus major triploids had similar growth rate to diploids at 10 months of age, but were smaller and presumed to be sterile at 18 months. Dicentrarchus labrax triploids showed inconsistent growth in relation to diploids and had lower Gonadal-Sematic Index (GSI).
	Improved growth and conversion efficiency in triploid rainbow trout, channel catfish, and at plaice flounder hybrids. Triploid Nile tilapia grew 66-90% better than diploids and showed decreased sex-dimorphism for body weight, but other studies found no advantage. Genotype by environment (GxE) interactions also influence performance.
	Triploid Pacific oysters show 13-51% growth improvement over diploids at 8-10 months of age and better marketability due to reduced gonads; triploid Sydney rock oysters showed 41% increase in body weight at 2.5 years.
	Polyploidization makes certain interspecific crosses viable.

232 for brown trout (Fjalestad, Moen and Gomez-Raya, 2003). Other fish species for which genetic databases and linkage maps are being created are tilapia (Kocher *et al.*, 1998) and channel catfish (Liu, 1999). The fish with the longest history of genetic alteration and improvement is the common carp (Balon, 1974). Breeding centres in Eastern Europe list over one hundred genetically distinct varieties, many of which are differentiated morphologically (Bakos and Gorda, 2001).

The Genetic Improvement of Farmed Tilapia (GIFT) programme increased the Nile tilapia growth rate by about 11 percent/generation (Eknath *et al.*, 1993) through selective breeding; the GIFT fish is now a registered trademark. Chromosome set manipulation has also been used to improve tilapia growth through the creation of all male tilapia or "genetically male tilapia" (GMT) (Mair *et al.*, 1995). Male tilapia grow faster than females and a single-sex population does not have the problem of unwanted reproduction.

Interspecific hybridization has been used to develop some animals that are useful for aquaculture, such as sunshine bass, which is a hybrid between white and striped bass, the bester, a popular sturgeon hybrid of beluga and starlet sturgeons, and red tilapia, which is produced by several crosses of various tilapia species (Bartley, Rana and Immink, 2001). In general, hybridization is not a good mechanism for creating stable varieties for production because the progeny may not be fertile, or when fertile, the second generation (F2) yields a group of fish with diverse phenotypes.

Gene transfer in fish is a technology that may have potential once environmental and human health issues are better understood by consumers. There are tremendous possibilities available to create new varieties, improve efficiencies and increase farming areas through genetic engineering. Advances in molecular genetics have allowed numerous useful genes to be identified and inserted into aquatic species (Table 5). At present, there are no transgenic aquatic species available to the consumer.

2.5 LOOKING AHEAD

Looking into the future aquaculture will grow rapidly and capture fisheries will level off. According to "The promise of the blue revolution" (*The Economist*, August 2003), aquaculture will provide most of the world's supply of fish by 2030. In many developed countries, inland food fisheries have been replaced by recreational fisheries, a trend that is also seen in some developing countries. To keep apace with human population growth, fishery production must increase if the same level of consumption of fish products enjoyed today is to be maintained. One strategy to provide additional food will be improved management of natural fisheries, taking into account genetic stock structure and the resilience and resistance that genetic resources give to natural populations. Another will be the further development of aquaculture, but this must be responsible development.

In the terrestrial agriculture sectors, species have been domesticated over millennia into diverse breeds. The fishery and aquaculture sectors use many more

TABLE 5
Examples of gene transfer involving aquatic species (FAO, 2000)

Species	Foreign gene	Desired effect and comments	Country
Atlantic salmon	AFP AFP salmon GH	Cold tolerance; increased growth and feed efficiency.	United States, Canada United States, Canada
Coho salmon	Chinook salmon GH + AFP	After 1 year, 10- to 30-fold growth increase.	Canada
Chinook salmon	AFP salmon GH	Increased growth and feed efficiency.	New Zealand
Rainbow trout	AFP salmon GH	Increased growth and feed efficiency.	United States, Canada
Cutthroat trout	Chinook salmon GH + AFP	Increased growth.	Canada
Tilapia	AFP salmon GH	Increased growth and feed efficiency; stable inheritance.	Canada, United Kingdom
Tilapia	Tilapia GH	Increased growth and stable inheritance.	Cuba
Tilapia	Modified tilapia insulin-producing gene	Production of human insulin for diabetics.	Canada
Salmon	Rainbow trout lysosome gene and flounder pleurocidin gene	Disease resistance, still in development.	United States, Canada
Striped bass	Insect genes	Disease resistance, still in early stages of research.	United States
Mud loach	Mud loach GH + mud loachand mouse promoter genes	Increased growth and feed efficiency; 2- to 30-fold increase in growth; inheritable transgene	China, Korea
Channel catfish	GH	33% growth improvement in culture conditions.	United States
Common carp	Salmon and human GH	150% growth improvement in culture conditions; improved disease resistance; tolerance of low oxygen level.	China, United States
Indian Major carps	Human GH	Increased growth.	India
Goldfish	GH AFP	Increased growth.	China
Abalone	Coho salmon GH + various promoters	Increased growth.	United States
Oysters	Coho salmon GH + various promoters	Increased growth.	United States
		Fish to other life forms	
Rabbit	Salmon calcitonin-producing gene	Calcitonin production to control calcium loss from bones.	United Kingdom
Strawberry and potatoes	AFP	Increased cold tolerance.	United Kingdom, Canada

Note: The development of transgenic organisms requires the insertion of the gene of interest and a promoter, which is the switch that controls expression of the gene.
AFP = anti-freeze protein gene (Arctic flatfish)
GH = growth hormone gene

species, but it has been suggested that there should be a reduction in this number and that these few domesticated species should be widely used throughout the world. This would mean an increase in the use of alien species and genotypes. At present, aquaculture is the primary reason for the deliberate movement of aquatic species, which have both good and bad impacts (FAO, 1999). Which model should the aquatic sector follow? Domestication of local species for local use, introduction of a few "good" species throughout the world, or better management of wild fisheries? Improved knowledge of the genetic diversity of aquatic species and how it functions in populations and ecosystems will help in evaluating these options.

Advances in biotechnology have and will continue to provide valuable information on aquatic diversity. Molecular analysis has demonstrated, however, that cows are genetically more similar to dolphins than to horses, but we do not think of putting cows in the sea or dolphins on the open range (paraphrased from Lewin, 1998). Biotechnology is one tool that can be used with others to help us develop, use and enjoy the tremendous aquatic diversity around us.

2.6 REFERENCES

Bakos, J. & S. Gorda. 2001. Genetic resources of common carp at the Fish Culture Research Institute, Szarvas, Hungary. *FAO Fisheries Technical Paper*, 417. Rome.

Balon, E.K. 1974. *Domestication of the common carp*, Cyprinus carpio L. Ontario, Canada, Royal Ontario Museum of Life Sciences Misc. Pub. 37.

Bartley, D.M. 1998. Genetics and breeding in aquaculture: current status and trends. In D.M. Bartley & B. Basurco, eds. *Genetics and Breeding of Mediterranean Aquaculture Species. Cahiers OPTIONS*, 34:13-30.

Bartley, D.M., Bentley, B., Brodziak, J., Gall, G., Gomulkiewicz, R. & Mangel, M. 1992. Geographical variation in the population genetic structure of Chinook salmon from California and Oregon. *Fishery Bulletin*, 90: 77-100.

Bartley, D.M. & Pullin, R.S.V. 1999. Aquatic genetic resources policy. In R.S.V.Pullin, D.M. Bartley & J. Kooiman, eds. *Towards policies for conservation and sustainable use of aquatic genetic resources*, pp. 1-16. ICLARM Conference Proceedings 59, Manila, Philippines.

Bartley, D.M., Rana, K. & Immink, A.J. 2001. Interspecific hybrids in aquaculture and fisheries. *Rev. Fisheries and Fish Biol.*, 10: 325-337.

Convention on Biological Diversity. 1992. (available at www.biodiv.org/welcome.aspx).

Eknath, A., Tayamen, M.M., Palada de Vera, M.S., Danting, J.C., Reyes, R.A., Dionisio, E.E., Capili, J.B., Bolivar, J.L., Abella, T.A., Circa, A.V., Bensten, H.B., Gjerde, B., Gjedrem, T. & Pullin, R.S.V. 1993. Genetic improvement of farmed tilapia: the growth performance of eight strains of *Orechromis niloticus* tested in different farm environments. *Aquaculture*, 111: 171-188.

FAO. 1999. *Impact of introductions on the conservation and sustainable use of aquatic biodiversity*, by D.M. Bartley & C.N. Casal. *FAO Aquaculture Newsletter*, 20: 15-19. Rome.

FAO. 2000. *State of the world fisheries and aquaculture 2000.* (available at www.fao.org/DOCREP/003/X8002E/X8002E00.htm).

FAO. 2002. State of world fishery resources: inland fisheries. *FI Circular*, 942. Rome.

FAO. 2005. *FishStat Plus.* Fisheries Department. Rome. (available at www.fao.org/fi/statist/fisoft/FISHPLUS.asp).

Fjalestad, K.T., Moen, T. & Gomez-Raya, L. 2003. Prospects for genetic technology in salmon breeding programmes. *Aquaculture Research*, 34: 397-406.

Government of Kenya. 2003. *Samaki News*, Vol. II, July, 2003, p. 33. Kenya Fisheries

Department.

Kocher, T.D., Lee, W.J., Sobolewska, H., Penman, D. & McAndrew, B. 1998. A genetic linkage map of a cichlid fish, the tilapia *(Oreochromis niloticus). Genetics*, 148: 1225-1232.

Lewin, R. 1998. Family feuds. *New Scientist*, January: 36-40.

Liu, Z. 1999. The catfish genetic linkage mapping, its current status and future prospects. *Plant and Animal Genome*, 7: 34.

Mair, G.C., Abucay, J.S., Beardmore, J.A. & Skibinski, D.O.F. 1995. Growth performance trials of genetically male tilapia (GMT) derived from YY-males in L: on station comparisons with mixed sex and sex reversed male populations. *Aquaculture*, 137: 313-342.

The Economist. 2003. *The promise of the blue revolution.* 9 August 2003:19-21.

3. Global overview of crop genetic resources

Brad Fraleigh

3.1 SUMMARY

Crop genetic resources activities include acquisition, conservation, maintenance, characterization and evaluation for use in the creation of new crop varieties and seed. The FAO Commission on Genetic Resources for Food and Agriculture (CGRFA) has launched a process to update the 1996 *Report on the State of the World's Plant Genetic Resources for Food in Agriculture (SoW-PGRFA)*, which remains an authoritative source of information and assessment.

3.2 INTRODUCTION

Presenting crop genetic resources at the global scale is an enormous task; this chapter will only attempt to provide an overview of findings and major global agreements, and present the status and progress made in certain critical activities.

Crop genetic resources include farmer's varieties and landraces, elite and special material and crop varieties developed by plant breeders and other researchers, wild and weedy relatives of crop plants, and wild plants harvested for food. They are used as raw material for the production of new crop varieties that respond to the needs of farmers and consumers. They provide insurance to meet future challenges posed by changes in the environment, diseases, and marketing opportunities, among others. Many crop cultivars also have significant cultural value for their holders.

Figure 6 illustrates the links between the conservation of plant genetic resources for food and agriculture (PGRFA) and its sustainable use as seed and propagating material, highlighting the key role of plant breeding. The system is farmer-centred as all the elements are integrated by farmers working in agro-ecosystems. Each component is related to the others, with germplasm flow and feedback loops connecting them.

Crop genetic resources systems typically acquire genetic resources, conserve their viability and genetic integrity, characterize diversity by evaluating its agronomic value, document all these activities, and facilitate their use by providing access to samples of material and associated information.

3.3 MAIN FINDINGS AND CHALLENGES

3.3.1 The first *Report on the State of the World's Plant Genetic Resources for Food and Agriculture*

In 1996, the first *SoW-PGRFA* Report was presented to 150 countries at the Fourth International Technical Conference on Plant Genetic Resources held in Leipzig, Germany (FAO, 1996a). Prepared under the auspices of FAO, it was welcomed as

FIGURE 6

Interrelationships between the conservation of crop genetic resources and their sustainable use in plant breeding and as seed and propagating material

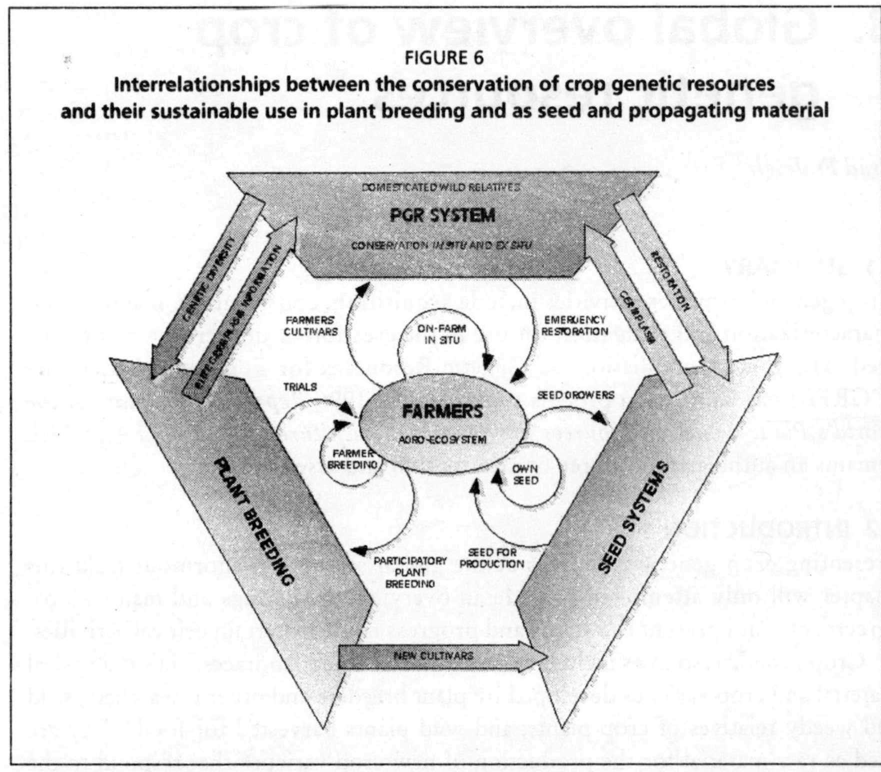

the first comprehensive evaluation in this domain. An extended version of the first *SoW-PGRFA Report* was published by FAO in 1997 (FAO, 1997). Although now almost ten years old, the first *SoW-PGRFA Report* remains the most widely cited source of information and assessment.

The first *SoW Report* pointed out that a small number of cereal crops provide a large proportion of total food requirements. However, when food energy supplies were analysed on a subregional level, a greater number and more types of crops emerged as significant. A substantial share of energy intake was also provided by meat, which is ultimately derived from forage and rangeland plants. For the most part, these plants were poorly collected, documented and exploited.

The first *SoW Report* provided summary information on many *ex situ* collections of major staple food crops. Over 1 300 gene banks were identified, preserving more than 5.5 million accessions, with wheat and rice accounting for 50 percent of the total. Continued gene flow between crops and their wild relatives in centres of origin underlined their importance as sources of new variability. Secondary centres of diversity were also very important. Various *in situ* approaches were identified: conservation of crop wild relatives and wild food plants; managed ecosystems such as rangelands; and conservation of traditional crop varieties on-farm and in home gardens. The report documented the facilitation of the use of

PGRFA by evaluating agronomic characteristics, pre-breeding, crop improvement and seed supply programmes.

Fifty-nine countries reported national committees on PGRFA but some lacked a national programme of any type. Almost 80 percent of these countries referred to lack of training as a serious constraint. The first *SoW-PGRFA Report* also documented the status of regional and international collaboration in PGRFA, access to genetic resources and benefit-sharing. A technical annex reported on the state of the art of methods to analyse and assess genetic diversity and vulnerability, *in situ* conservation and methods for utilization of PGRFA through plant breeding.

3.3.2 Challenges

Conserving and using PGRFA effectively requires policies, knowledge and action. Crop genetic resources are at the intersection of agriculture, food security, environment and trade. A series of relevant global agreements affects their management, including the World Food Summit Plan of Action, the CBD, and global agreements on intellectual property rights.

The exchange of crop genetic resources, from which everyone will eventually benefit, has been a reality since the beginning of agriculture. Our growing world population will only be fed if the widest possible range of genetic diversity can be draw upon. As a result, countries and regions are interdependent. The world greatly depends on crops that originated elsewhere for food and agriculture. No country can hope to do well with its own resources alone. Finally, it is a fact that PGRFA were developed and are maintained by humans - by farmers within their farming systems and by plant breeders and researchers. Unless diversity is conserved in gene banks, it is often lost when farming systems die or when scientific programmes come to an end.

At the scientific and technical levels, challenges are posed by genetic erosion, genetic vulnerability and utilization. Value in crop genetic resources lies at the intra-specific level, in the diversity *within* a crop's gene pool. The first *SoW-PGRFA Report* defined genetic erosion as "the loss of genetic diversity, including the loss of individual genes, and the loss of particular combinations of genes (i.e. of gene-complexes) such as those manifested in locally adapted landraces" (FAO, 1997). There is still no scientific consensus on how to measure genetic diversity in a crop's gene pool. Similarly, there is no consensus on the optimal balance of *in situ* and *ex situ* conservation methods to combat genetic erosion. The first *SoW-PGRFA Report* identified measures to strengthen *ex situ* conservation, such as "the development of low-cost conservation technologies, and in particular, technologies for non-orthodox seeded and vegetatively propagated plants including *in vitro* methods and cryopreservation" (FAO, 1996a).

Genetic vulnerability has been described as "the condition that results when a widely planted crop is uniformly susceptible to a pest, pathogen or environmental hazard as a result of its genetic constitution, thereby creating a potential for widespread crop losses" (National Academy of Sciences, 1972, cited in FAO, 1996a).

TABLE 6
Some biotechnologies used in the conservation and use of crop genetic resources

Objective	Activity	Technology
Conservation	clonal gene banks	*in vitro* conservation
		cryopreservation
	plant health	ELISA diagnostics
		virus treatment
	identification	genetic fingerprinting
Characterization	protein analysis	isoenymes
	DNA analysis	RAPD, RFLP, microsatellites, gene arrays, sequencing
Evaluation	genetic markers	QTL mapping, genetic maps
Enhancement	wide crossing	embryo rescue
	transgenes	transformation, gene expression

Genetic vulnerability pertains to the level of the crop genetic diversity actually being used. One of its main causes is the widespread replacement, since the 1950s, of genetically diverse traditional varieties by varieties with more homogenous genetic make-up. The first *SoW-PGRFA Report* noted that "there is no comprehensive or coordinated system for monitoring uniformity in agricultural species, and methodological tools which might help assess related genetic vulnerability have not been adequately developed" (FAO, 1996a). Over the past decade, farmers' reasons for maintaining or changing the varieties they grow have only just begun to be understood.

Conservation of the diversity of crop genetic resources is not enough: their potential uses and values need to be understood by characterizing, evaluating and documenting them. Methods still need to be developed to improve and facilitate productive utilization. The first *SoW-PGRFA Report* noted that "a variety of plant breeding and biotechnological techniques, which often differ in technical complexity and cost, may be used in crop improvement... Biotechnological methods are now increasingly available to facilitate wide crosses thus allowing the introduction of the desired genes" (FAO, 1996a). It also noted that "while a number of countries have initiated crop improvement programmes based on new biotechnologies, not all countries have the capacity to use such technologies" (FAO, 1997). There are biotechnologies relevant for every objective of a crop genetic resources system. Table 6 provides some examples.

Another set of challenges is posed for taking action: Is there sufficient capacity for this? How can capacity be built where it is currently inadequate? How can the necessary cooperation be organized among countries and among disciplines, particularly in order to link conservation and use of crop genetic resources? How can the resources needed to address these issues be mobilized?

3.4 THE INTERNATIONAL TREATY ON PLANT GENETIC RESOURCES FOR FOOD AND AGRICULTURE

The International Treaty on Plant Genetic Resources for Food and Agriculture - a new, legally binding international instrument - provides a global framework to respond to such challenges. The FAO Conference adopted the International Treaty by consensus on 3 November 2001, and it entered into force on 29 June 2004 (FAO, 2001).

The Treaty provides an internationally agreed framework for the conservation and sustainable use of all PGRFA. Its objectives are "the conservation and sustainable use of PGRFA and the fair and equitable sharing of the benefits arising out of their use, in harmony with the CBD, for sustainable agriculture and food security" (Article 1). All of the Treaty's provisions are consistent with the Convention.

In Article 9, the Treaty recognizes

> "the enormous contribution that the local and indigenous communities and farmers of all regions of the world, particularly those in the centres of origin and crop diversity, have made and will continue to make for the conservation and development of plant genetic resources which constitute the basis of food and agriculture production throughout the world".

A cornerstone of the Treaty is a Multilateral System of Access and Benefit-Sharing (Part IV: art. 10-13). The Multilateral System covers a list of crops established according to criteria of food security and interdependence. These crops provide about 80 percent of the food derived from plants. Governments will bring into the Multilateral System all the genetic resources that are under their management and control and in the public domain. They will encourage other holders of PGRFA within their country to place the resources they hold into the Multilateral System, including those kept by the Consultative Group on International Agricultural Research (CGIAR).

The Treaty recognizes that facilitated access to these PGRFA is in itself a major benefit. This will ultimately benefit consumers by providing a stream of improved and varied agricultural products. The Treaty also identifies and makes provision for a wide range of other forms of benefit-sharing, including information exchange, access to and transfer of technology, capacity-building, and sharing the financial and other benefits of commercialization.

3.5 FAO COMMISSION ON GENETIC RESOURCES FOR FOOD AND AGRICULTURE

Since 1983, the world has a permanent forum where governments discuss and negotiate matters relevant to genetic resources for food and agriculture. The main objectives of the FAO Commission on Genetic Resources for Food and Agriculture (CGRFA) are to ensure the conservation and sustainable utilization of genetic resources for food and agriculture (not only plants), as well the fair and equitable sharing of benefits derived from their use, for present and future generations. The Commission aims to reach international consensus on areas of global interest. At present, 167 countries and the European Community are members.

Since its establishment, the Commission has overseen the development of a series of international instruments and global mechanisms (see overview in FAO, 2004b).

FIGURE 7
10-15 year cycle of information-gathering, planning and action in global crop genetic resources

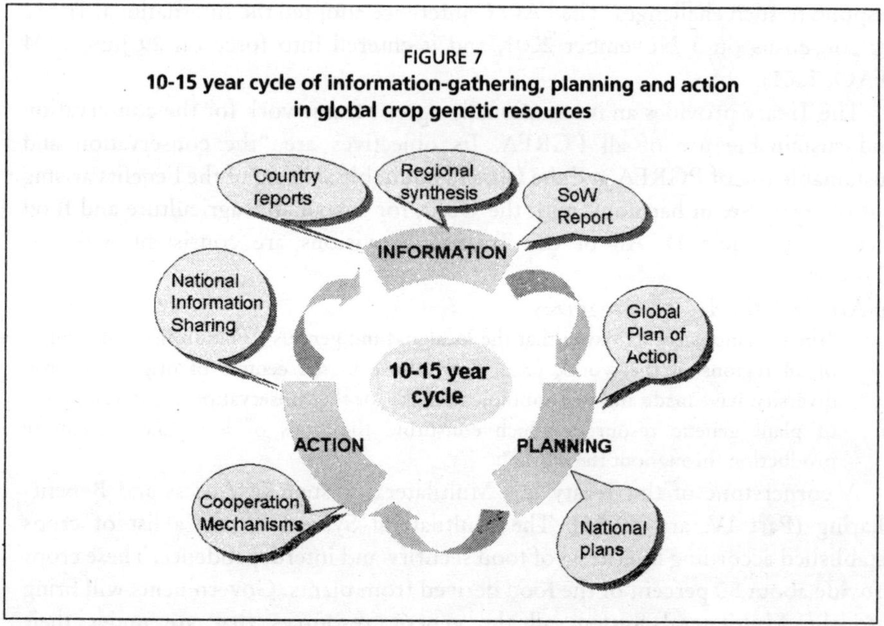

Many of these were integrated into the International Treaty as "supporting components". The first *SoW Report* was the basis of the rolling *Global Plan of Action* for the Conservation and Sustainable Utilization of Plant Genetic Resources for Food and Agriculture (the *Global Plan of Action*). These two important components of the Commission's activities are related according to the 10-15 year cycle shown in Figure 7.

3.5.1 *Global Plan of Action*

The *Global Plan of Action* provides a comprehensive scientific and technical framework for international and national action at the global level. It was adopted by 150 countries in 1996 through the Leipzig Declaration and was endorsed by the World Food Summit Plan of Action and the CBD (FAO, 1996b).

The *Global Plan of Action* presents 20 Priority Activity Areas (PAA) organized in four main groups: *in situ* conservation and development; *ex situ* conservation; utilization of plant genetic resources; and institutions and capacity-building. Each PAA includes a section on Research/Technology where areas of scientific, methodological, or technological research or action relevant to the implementation of the priority activity are identified, including technology development and transfer.

The *Global Plan of Action* identifies a role for biotechnologies in PAA 11: "Promoting sustainable agriculture through diversification of crop production and broader diversity in crops". One of its objectives is: "to promote the goal of higher levels of genetic diversity consistent with productivity increase and agronomic needs, including in crop production, plant breeding and biotechnological research and development settings". Regardy capacity: "governments, and their national agricultural research systems, supported by the International Agricultural Research Centres, and other research and extension organizations should: ... (e) make use of modern biotechnological techniques as feasible, to facilitate broadening of the genetic base of crops".

Biotechnologies have a role in realizing other priority activities of the *Global Plan of Action*, in particular, but not exclusively PAA 5: "Sustaining existing *ex situ* collections"; PAA 8: "Expanding *ex situ* conservation activities"; PAA 9: "Expanding characterization, evaluation and number of core collections to facilitate use"; and PAA 10: "Increasing genetic enhancement and base-broadening efforts".

3.6 CURRENT GLOBAL INITIATIVES

3.6.1 The Leipzig Conference took place in 1996. What has happened more recently?

The Leipzig Conference agreed that overall progress in the implementation of the *Global Plan of Action* and the related follow-up processes would be monitored and guided by national governments through the Commission. Global surveys to monitor implementation of the *Global Plan of Action* were conducted in 1998, 2000, 2002 and 2004 using an open-ended narrative format with some tabular, yes/no and multiple choice answers. The degree of participation was relatively high in the first two surveys and somewhat lower in 2004.

A country progress report presented to the Commission in November 2004 (FAO, 2004a) provides an overview of the changes that occurred during 2001-2003 in the 77 countries that participated in the 2004 survey. The use of biotechnologies was reported in respect of PAA 8: "Expanding *ex situ* conservation activities". Thirty percent of the countries published information on improved methodologies for *ex situ* conservation; micro-propagation was the new technology most widely applied. Examples of activities reported included the establishment of a cryo-bank to conserve vegetative tips of potato and hops in the Czech Republic, and the development of low-cost seed preparation and storage techniques for sorghum germplasm in the Sudan.

In respect to PAA 9: "Expanding characterization, evaluation and number of core collections to facilitate use", greatly increased use of molecular methods for characterization was reported in some regions. Ninety percent of European countries and 77 percent of countries in Asia and the Pacific now employ these. Only 28 percent of the reporting African countries stated they used molecular methods for this activity area, although this percentage tripled since the previous survey in 2001.

3.6.2 A new approach to monitor the implementation of the *Global Plan of Action*

The Commission identified three limits to the global survey: lack of detailed information, which restricted the depth of analysis, lack of quantitative information, and an inadequate range of sources of information. Based on the guidance of the Commission, a new approach for monitoring implementation was developed starting in 2001.

Among the main considerations in designing the new monitoring approach were that it should directly benefit national programmes and be as participatory as possible. The new approach (FAO, 2005a) relies on four main components:

(i) a list of indicators for monitoring the implementation of all PAAs at the country level;
(ii) a reporting format, which is a structured questionnaire based on these indicators;
(iii) a computer application to facilitate and simplify recording, processing, analysis and sharing information;
(iv) guidelines for initiating and coordinating this process, including the involvement of stakeholders and the establishment of a national information-sharing mechanism.

In 2002, the Commission highlighted the importance of monitoring the implementation of the *Global Plan of Action* through a country-driven and flexible system, while ensuring the necessary level of standardization. Representatives of countries participating in pilot testing found that the new approach entailed more work on their part, but that it resulted in much better information at the national level. Stakeholder participation was greatly enhanced and national information-sharing mechanisms were made accessible through Web sites managed by countries (see www.pgrfa.org) and compact discs distributed by the national focal points.

In November 2004, the Commission supported the application of the new monitoring approach to all countries, in view of the integration of the monitoring activities with the preparation of the second *SoW-PGRFA Report*. The results of the new approach are already significantly updating the knowledge of the state of the world's crop genetic resources at the national level.

3.6.3 Development of the second Report on the State of the World's Plant Genetic Resources for Food and Agriculture

In 2002, the Commission agreed to the preparation of the second *SoW-PGRFA Report*, giving priority to updating the first report and emphasizing the changes that occurred since its publication. The outline for the second *SoW-PGRFA Report* is shown in Figure 8.

The Commission also approved a list of thematic background studies (Figure 9), which will address issues that were not treated, or treated only superficially in the first *SoW-PGRFA Report*. There are biotechnology considerations present in each of the thematic background studies; studies B, C, D and J are of particular relevance.

In November, 2004, the Commission adopted steps for preparing the second *SoW-PGRFA Report*. These follow a bottom-up approach: first, preparation of country reports, then regional synthesis, followed by global synthesis. The

FIGURE 8

Outline for the second Report on the State of the World's Plant Genetic Resources for Food and Agriculture

1. The state of diversity
2. The state of *in situ* management
3. The state of *ex situ* conservation
4. The state of use
5. The state of national programmes, training needs and legislation
6. The state of regional and international collaboration
7. Access to plant genetic resources, sharing of benefits derived from their use, and farmers' rights
8. The contribution of PGRFA management to food security and sustainable development

Annex 1 The state of the art: methodologies and technologies for the identification, conservation and use of PGRFA

Annex 2 The state of diversity of major crops and other PGRFA

Table 1 Status by country of national legislation, programmes and activities for PGRFA

Commission confirmed that the preparatory process should be fully integrated into the process of monitoring the implementation of the *Global Plan of Action* in order to minimize the reporting burden. It called upon donor countries and international organizations to assist by providing the financial resources required for the full participation of all countries.

FAO and International Plant Genetic Resources Institute (IPGRI) presented *Guidelines for Country Reports* (FAO, 2005b) to the Commission's Working Group on Plant Genetic Resources in October 2005. The guidelines point out that countries should assess the current contribution of PGRFA, the state of PGRFA in the country and their role in production systems. This includes associated biodiversity, the factors driving change, and how the contribution of PGRFA can

FIGURE 9

List of thematic background studies contributing to the second *SoW Report*

A Plant genetic resources of forage crops, pasture and rangeland.

B The conservation of crop wild relatives.

C Indicators of genetic diversity, genetic erosion and genetic vulnerability.

D Methodologies and capacities for crop improvement; the use of PGRFA in base-broadening and crop improvement, including new approaches to plant breeding and new biotechnologies.

E Seed security for food security: the management of plant genetic resources in seed systems.

F The contribution of plant genetic resources to health and dietary diversity.

G Managing plant genetic resources in the agro-ecosystem; global change, crop-associated biodiversity and ecosystem services.

H Interactions between plant and animal genetic resources, and opportunities for synergy in their management.

I The impact of national, regional and global agricultural policies and agreements on conservation and use of PGRFA.

J Biosafety and biosecurity issues related to the conservation and sustainable utilization of PGRFA.

be enhanced, identifying opportunities and obstacles as well as strategies to realize the opportunities and overcome any obstacles.

Questions and tables generated by monitoring implementation of the *Global Plan of Action* are cross-referenced under each of the relevant chapters of the *Guidelines for Country Reports.* The monitoring provides data and information - the country reports analyse and synthesize. The Guidelines for the Country Reports encourage countries to take a strategic and forward-looking approach, which should prove useful for national planning. Recommendations are provided on how to use a participative approach in preparing country reports.

The Commission requested that the second *SoW-PGRFA Report* be submitted for adoption to its Twelfth Regular Session in late 2008. In order to meet this deadline, all countries will need to monitor their implementation of the *Global Plan of Action* by April 2007 and prepare their country reports by the end of June 2007. Regional discussions would take place during the latter half of 2007, followed by drafting and discussion of the final Draft Report during the first part of 2008.

In summary, tools are now available to monitor implementation of the comprehensive *Global Plan of Action.* They allow for greatly enhanced involvement of stakeholders and culminate in the creation of national information-sharing mechanisms. Since the *Global Plan of Action* involves biotechnologies, monitoring its implementation includes taking stock of their use over the past ten years and their potential further use. The guidelines integrate the preparation of country reports with the monitoring. The country reports provide an opportunity for countries to strategize their future use of new biotechnologies. These activities will lead to the elaboration of a second authoritative Report on the State of the World's Plant Genetic Resources for Food and Agriculture.

3.7 CONCLUSIONS

Policies and forums support the conservation and sustainable use of crop genetic resources at the global level, in particular the International Treaty on Plant Genetic Resources for Food and Agriculture and the CGRFA. There are also global instruments and institutions such as the *Global Plan of Action*, Reports on the State of the World's PGRFA, the Consultative Group on International Agricultural Research and others. Activities are taking place at many levels.

Crop genetic resources are managed at the intersection of agriculture, food security, environment and trade. Genetic erosion and genetic vulnerability continue to threaten crop diversity, but they can be overcome with good science, adequate resources, cooperation and political will.

3.8 REFERENCES

FAO. 1996a. First *Report on the state of the world's plant genetic resources for food and agriculture* (first *SoW-PGRFA Report*), prepared for the International Technical Conference on Plant Genetic Resources, Leipzig, Germany, 17-23 June 1996, Rome. (available at www.fao.org/WAICENT/FAOINFO/AGRICULT/AGP/AGPS/Pgrfa/pdf/swrshr_e.pdf).

FAO. 1996b. *Global Plan of Action* for the conservation and sustainable utilization of plant genetic resources for food and agriculture and the Leipzig Declaration, adopted by the International Technical Conference on Plant Genetic Resources, Leipzig, Germany 17-23 June 1996. (available at www.fao.org/WAICENT/FAOINFO/AGRICULT/AGP/AGPS/GpaEN/gpatoc.htm).

FAO. 1997. First *Report on the state of the world's plant genetic resources for food and agriculture*, extended version (first *SoW Report*, extended version), Rome. (available at www.fao.org/WAICENT/FAOINFO/AGRICULT/AGP/AGPS/pgrfa/pdf/swrfull.pdf).

FAO. 2001. International Treaty on Plant Genetic Resources for Food and Agriculture. (available at www.fao.org/ag/cgrfa/itpgr.htm).

FAO. 2004a. Country progress report on the implementation of the *Global Plan of Action* for the conservation and sustainable utilization of plant genetic resources for food and agriculture. Document CGRFA-10/04/Inf.6. Commission on Genetic Resources for Food and Agriculture, 10th Regular Session, 8-12 November 2004. (available at ftp://ext-ftp.fao.org/ag/cgrfa/cgrfa10/r10i6e.pdf).

FAO. 2004b. Overview of the FAO global system for the conservation and sustainable utilization of plant genetic resources for food and agriculture and its potential contribution to the implementation of the International Treaty on Plant Genetic Resources for Food and Agriculture. 10th Regular Session, 8-12 November 2004. Commission on Genetic Resources for Food and Agriculture. (available at ftp://ext-ftp.fao.org/ag/cgrfa/cgrfa10/r10w3e.pdf).

FAO. 2005a. Monitoring the implementation of the *Global Plan of Action* and preparation of the second Report on the state of the world's plant genetic resources for food and agriculture. Document CGRFA/WG-PGR-3/05/3. Intergovernmental Working Group on Plant Genetic Resources of the Commission on Genetic Resources for Food and Agriculture. 2005. 3rd Regular Session, 26-28 October 2005. (available at www.fao.org/waicent/FaoInfo/Agricult/AGP/AGPS/pgr/ITWG3rd/pdf/p3w3E.pdf).

FAO. 2005b. Preparation of the second report on the state of the world's plant genetic resources for food and agriculture: guidelines for country reports. Document CGRFA/WG-PGR-3/05/Inf.5. Intergovernmental Working Group on Plant Genetic Resources of the Commission on Genetic Resources for Food and Agriculture. 3rd Regular Session, 26-28 October 2005. (available at www.fao.org/waicent/FaoInfo/Agricult/AGP/AGPS/pgr/ITWG3rd/docsp1.htm.

National Academy of Sciences. 1972. *Genetic vulnerability of major crops.* Washington, USA. Cited in the first *SoW Report.*

4. Efforts towards assessing the global status of forest genetic resources

Pierre Sigaud

4.1 SUMMARY

Developments and applications of forest genetic resources (FGR) are becoming wider in coverage, more specialized and more accessible throughout the world. Efforts to assess, monitor and report on the state of the world's forests and biological diversity have so far only partially addressed the genetic level, although several frameworks exist for a global assessment of FGR. Such an assessment presents special challenges. This chapter reviews past and ongoing efforts to review the status and trends of forest tree genetic resources, and provides options for future developments.

4.2 INTRODUCTION

Forests are the single most important repositories of terrestrial biological diversity. Forest trees and woody plants have developed complex mechanisms to maintain high levels of genetic diversity. This genetic variation, between and within species, is mainly conserved on site (*in situ*), and allows tree and shrub species to react against variations in the environment, pests, diseases and climatic change. It provides the foundation for future evolution, selection and breeding. In addition, at different levels, it supports the aesthetic, ethical and spiritual values given to forests and trees by human beings.

Although anthropological influences are found in most of the world's forests, trees have only been partly domesticated in the past half-century. In effect, very few trees are more than one or two generations removed from their wild congeners, unlike most agricultural crop plants. FGR have traditionally been associated with the provision of forest reproductive material, through tree selection and improvement. In the past two decades, the FGR field has increasingly integrated genetic conservation and management considerations. The scope of FGR is now rapidly expanding to include advances in biotechnology, biosecurity (management of biological risks) and legal developments concerning access rights and benefit-sharing.

The FGR field is mainly driven by three technical sectors, namely, forestry, the environment and agriculture. (Policies related to FGR trade and technological and regulatory developments are often dealt with under crop genetic resources.) The relative importance of these three driving sectors varies over space and time. (Figure 10).

FGR are the subject of active scientific and technical research worldwide. However, information on applications of the research to forestry or biodiversity

FIGURE 10
The main sectors driving policy, regulatory and technical developments in FGR

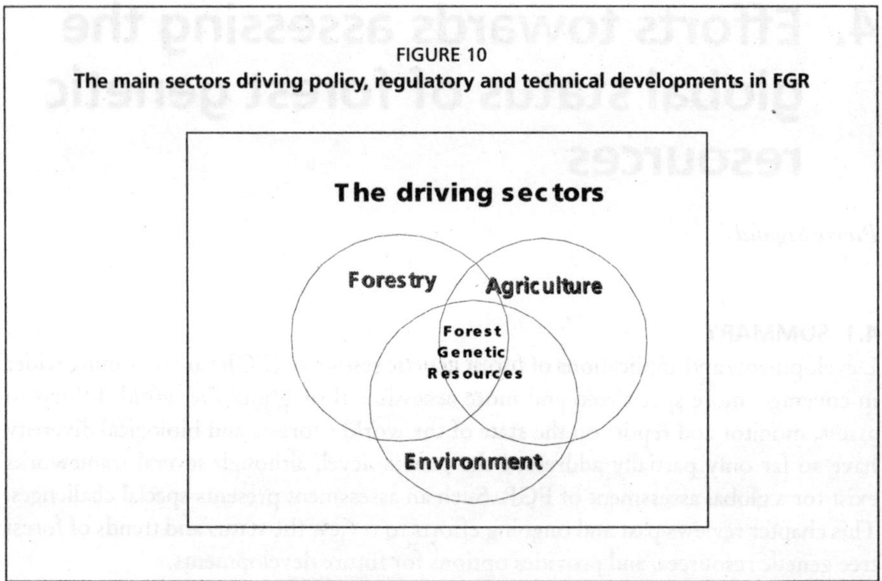

management is generally patchy and inconsistent. FGR research can find broad application in:

(i) *Commercial plantations*. Forest plantations tend to follow an agricultural model, with significant private sector investment, a focus on just a few species, best sites and short rotations for the provision of wood for industrial purposes. Commercial plantations cover 3 percent of the global forest area and already provide 35 percent of the global industrial wood. Their share of industrial wood products is expected to increase in the next decades.

(ii) Specific niches, including tree domestication for agro-forestry, applications for species survival, soil conservation, habitat restoration and urban and amenity forestry.

(iii) Natural and semi-natural forests. Significant research efforts use biotechnology but few applications have so far been reported. Research areas include ecosystem functioning, tree biology and patterns of genetic variation. Integrating FGR in management policies and plans will be a global challenge.

4.3 GLOBAL ASSESSMENT/REPORTING OF RELEVANCE TO FOREST GENETIC RESOURCES

4.3.1 Global assessment of crop genetic resources

There is no forestry equivalent to the *Global Plan of Action* for the Conservation and Use of Plant Genetic Resources for Food and Agriculture that focuses on agricultural crops. The *Global Plan* makes reference to wild relatives of cultivated plants, often found in forest ecosystems, and to domesticated trees such as fruit trees and rubber, but explicitly excludes forest tree genetic resources.

The first *SoW-PGRFA Report*, published in 1997 by FAO is currently being revised and updated. The 1997 edition makes several references to forest tree genetic diversity and to methodologies developed for *in situ* conservation. In the second version, planned for 2006, several developments related to trade, biotechnology, genetic modification, and regulatory frameworks on biosecurity and biosafety are likely to apply to forestry.

4.3.2 Global assessment, monitoring and reporting on forests

The high-level international forest policy dialogue initiated after the United Nations Conference on Environment and Development (UNCED), and now continued through the United Nations Forum on Forests (UNFF), has generated more than 270 proposals for action towards sustainable forest management. Few references are made to FGR (mainly on intellectual property issues), and there is no reference to forest genetic resources assessment. However, the Collaborative Partnership on Forests, gathering 14 forest-related international organizations, institutions and convention secretariats to support the work of the UNFF, has set up a portal on streamlining forest-related reporting. The portal is designed to help users access country reports, some of which address genetic diversity.

The FAO Forest Resources Assessment (FRA) programme regularly gathers, compiles and publishes global statistics and analyses on forest cover and related fields. The FRA 2000 made a first attempt to review information on forest biological diversity and its links with FRA. While useful indications were given on the extent of information available by species groups (Table 7), it was recognized that assessing biological diversity at the global level represents a major challenge. FRA 2000 highlighted that no single, objective measure of biological diversity could be used, but only complementary measures appropriate for specific and restricted purposes. Equally important, no agreed methodology was identified for linking changes in forest area, structure and composition to their impacts on forest landscapes, species, populations and genes, especially when information was

TABLE 7
Data availability by species group in Forest Resources Assessment 2000

Group	All species occurring in country				Forest-occurring species			
	All species		Endemic species		All species		Endemic species	
	Total	Endangered	Total	Endangered	Total	Endangered	Total	Endangered
Ferns	Good	Good	Limited	Good	Good	Good	Limited	Good
Palms	Good	Good	Good	Good	Good	Good	Good	Good
Trees	No data	Good1/	Limited	Good	No data	Good1/	Limited	Good
Amphibia	Good	Good	Partial	Good	No data	No data	No data	Good
Reptiles	Good	Good	Partial	Good	No data	No data	No data	Good
Birds	Good	Good	Partial	Good	No data	No data	No data	Good
Mammals	Good	Good	Partial	Good	No data	No data	No data	Good

1/ for most countries
Source: FAO, 2001

aggregated at the global level. More generally, the introduction of complex and intangible values, such as biological diversity, raises the need for a review of concepts and themes to be covered in global forest assessments.

The next issue of FRA, planned for release in 2006, will include country-based information of relevance to biodiversity, including forest extent, categories and characteristics, tree species occurrence, and tree species abundance, composition and vulnerability.

The *State of the Art Report on the Research on Forest Tree Genetic Diversity* was prepared for the XXI International Union of Forest Research Organizations (IUFRO) World Congress by the IUFRO Task Force on Management and Conservation of FGR in 2000. The report concluded that, in general, the state of scientific knowledge on the importance of FGR in various research areas, and in particular in forest management, was far from satisfactory. Conservation and utilization of gene resources were reported as being perceived as mainly biological and ecological issues, with limited linkage to policy and land use, economy and other issues.

4.3.3 Global reporting on biological diversity

The legally binding Convention on Biological Diversity (CBD), which entered into force in 1993, adopted an expanded work programme on forests at the Sixth Conference of the Parties in 2002. The programme (VI/22) makes specific reference to the genetic level and includes activities to: (i) develop, harmonize and assess the diversity of FGR; (ii) provide guidance for countries to assess the state of their FGR; and (iii) develop a holistic framework for the conservation and management of FGR at national, subregional and global levels. The formal inclusion of forest genetic diversity in a work programme of the CBD provides an important vehicle for national institutions to further justify and strengthen activities on FGR assessment, monitoring and reporting.

A preliminary assessment of the status, trends and identification of options for the conservation and sustainable use of forest biological diversity was carried out under the CBD in 1999. Information was used in the *Global Biodiversity Outlook* compiled by United Nations Environment Programme (UNEP)/CBD in 2001. The genetic level was addressed, mainly through qualitative statements and case studies.

The work of the Convention has more recently focused on the development of targets and indicators for assessing progress in the various work programmes. Some targets are of relevance to genetic resources. In 2002, for example, the target of "70 percent of the genetic diversity of crops and other major socio-economically valuable plant species conserved, and associated indigenous and local knowledge maintained" was adopted for the Global Strategy for Plant Conservation. Specific targets for forest biological diversity, which will include genetic-level indicators, are still under development.

Efforts by the CBD to review trends in genetic diversity of domesticated animals, cultivated plants and fish species of major socio-economic importance concluded in 2004 that "[t]here are very rarely many, if any, quantitative data on population size changes upon which to monitor variations in the genetic diversity of forest tree species, at least in tropical areas, which account for 80 percent of the world total forest tree species." Other reviews have recognized that most of the approaches to genetic diversity monitoring have problems in being too complex or not sufficiently representative of the global status, or involve too many measures. A major difficulty is that the analytical work needed to combine different measures into one or two simple and understandable indicators has yet to be done.

4.4 ASSESSMENT, MONITORING AND REPORTING AT REGIONAL AND NATIONAL LEVELS

4.4.1 Criteria and indicators for sustainable forest management

Criteria define the concept and the main aspects of sustainable forest management as well as the related values. Indicators are quantitative and/or qualitative attributes of each criterion and are used to assess the current status and to monitor trends over time. Some 150 countries formally adhere to one of nine major inter-governmental criteria and indicators processes. Biological diversity indicators in general, and genetic diversity indicators in particular, have presented difficulties for all processes. Some processes assign more importance to quantitative indicators, while others have taken a more pragmatic approach and specified qualitative indicators (Table 8).

A study conducted for FAO in 2002 concluded that the issue of genetic diversity was not well addressed in any process, except perhaps in the pan-European initiative. Many indicators of genetic diversity were not effective or lacked practicality, and their relevance to sustainable forest management was tenuous. Further development and testing of different surrogate attributes were found necessary, but the impetus for this appeared to be waning.

TABLE 8
Themes under which forest genetic resources issues are addressed according to criteria and indicators processes; broad categories of indicators assessed

Process	Theme	Instrument	Species	Population
African Timber Organization	Biodiversity		√	
Asian Dry Forests	Biodiversity	√		
African Dry Zones	Biodiversity	√	√	√
International Tropical Timber Organization (ITTO)	Biodiversity	√		
Lepaterique (Central America)	Environment		√	
Montreal (temperate and boreal forests)	Biodiversity		√	√
Near East	Biodiversity		√	√
Pan-European	Biodiversity	√		√
Tarapoto (South America)	Biodiversity	√		

TABLE 9
Workshops on forest tree genetic resources supported by FAO, IPGRI and DFSC from 1995 to 2003

Eco-region	No. of countries	Country status	Priority species	Regional summary	Regional action plan
Temperate North America (1995)	3		+	+	+
Boreal Forests (1995)	20		+	+	+
Sahelian Africa (1998)	15	+	+	+	+
Pacific Islands (1999)	18	*	+	+	+
Eastern, Southern Africa (2000)	9	+	+	+	+
Southeast Asia (2001)	8	+	+	+	+
Central America (2002)	9	+	+	+	+
Central Africa (2003)	6	*	+	+	*
South Asia (2003)	13	+	+	+	+

(+ indicates achievement; * indicates work in progress)

4.4.2 Regional workshops on forest and tree genetic resources

A number of workshops on FGR have been convened with the support of FAO, the IPGRI, the Danida Forest Seed Centre (DFSC), Denmark, and many other organizations (Table 9). Regional workshops have supported the development of national status reports and regional action plans for conservation and sustainable use of FGR. During the process, methodologies for assessing the state of forest tree genetic diversity at the country level were developed by local experts to describe the genetic management of important species. In most regions covered, country-based status reports have been prepared and synthesized in regional syntheses. Summarized information on species management has partly been compiled in the FAO information system on FGR, REFORGEN.

The above information was gathered using methodologies developed with a regional objective. In some areas, countries have focused on native tree species, excluding introduced trees. Patchy data structure and data quality control represent major challenges for a global use of the dataset in a global FGR assessment.

Many other organizations have established information systems of relevance to forest/tree genetic resources, with different focus and purposes, including:

- the Tree Seed Supplier Directory and the Agroforestree Database (World Agroforestry Centre [ICRAF]);
- the Red List of Threatened Species (The World Conservation Union [IUCN]);
- the Tree Conservation Database on Endangered Species (World Conservation Monitoring Centre [WCMC]);
- the Forestry Compendium (CABI);
- the database on approved forest materials, by species and country (Organisation for Economic Co-operation and Development [OECD]);
- the collection of forest tree genome databases, Dendrome (University of California);
- the World Directory of Forest Geneticists and Tree Breeders (North American Forest Commission [NAFC] and IUFRO).

In summary, although rationales and frameworks exist for a global assessment of FGR, such an assessment is still to be undertaken.

4.5 THE WAY FORWARD

In 2003, the FAO Panel of Experts on Forest Gene Resources recognized the relevance and discussed the feasibility of a global FGR assessment. As a source of official, harmonized and validated information, the assessment could guide national and international processes and institutions towards sustainable forest and biodiversity management. Further discussions highlighted that such assessment should include thematic issues and case studies, trends and options for the future, rather than being a mere assembly of baseline statistics. Follow-up action towards a global FGR assessment is currently being discussed.

It is generally acknowledged that the complexity of genetic diversity at the global level must be expressed in a simplified, uniform and easily understood set of variables that represent the major values. Such variables must by necessity be based on generalizations that use indirect (surrogate) measures, typically indicators based on the general (qualitative) condition of the resource.

Prerequisites to the assessment include clarification of the concept (FGR or forest genetic diversity?); harmonization of definitions to be used globally; scope of the assessment; development of agreed methodology; identification of value (utility) given to genetic resources by stakeholders/users; identification and ranking of important (priority) tree species (reference is often made to three groups – species of current socio-economic importance, species with clear potential or future value, and species of unknown value given present knowledge and technology); classification of management practices and threats; actual attributes and potential use of trees; and clarification of what can reasonably be done in a given time frame (four to five years).

The assessment will likely be constrained by the availability of country-based statistics and rely on complementary thematic studies, including on species/genera of global importance. The assessment will also emphasize status and trends in major FGR issues. Very basic indicators on genetic diversity (for example, distribution of major tree species in different ecological zones) could then be considered. Both the genetic resources and their conservation and use are essentially dynamic; therefore underlying genetic processes may also have to be considered. Priority will be given to carefully defining the end result and using information sources and analyses in a scientifically sound and transparent way.

4.6 CONCLUDING REMARKS

Global priorities in FGR have changed from an early focus on seed, species and provenance research of a few timber species in the 1960s and 1970s, to the wider management of genetic diversity of a range of trees and shrubs for a number of purposes and end uses in a variety of contexts. These developments are often led by and under the umbrella of agreements in other sectors, including trade and

BOX 1

Themes that may be considered in a global forest genetic resources assessment

Concept, scope, definitions
Estimators and surrogates of forest tree genetic diversity
Important forest tree genera, species, provenances and varieties
State of conservation *in situ*, *ex situ*
State of use, selection and breeding programmes
Seed and reproductive materials: pathways, supply and demand
Legal issues, property rights, access, benefit sharing, material transfer agreements
State of biotechnology research and application
Economic rationales in conventional and biotechnology-based programmes
Biosecurity issues: invasive alien tree species, genetic pollution, GM tree deployment
Research and education, national/public/private capacities status and trends
Institutional framework, international and national programmes
State of diversity and vulnerability of major genera/species
Other thematic/case studies

economics, agriculture and the environment, rather than being integral parts of forestry-led initiatives. Some developments may be anticipated to some extent. Globally, increased movement of people, goods, services, information and know-how contributes to a constant change in demands on and value of forests, wood and non-wood products and environmental services, and to shifts in the boundaries, and priorities of the FGR sector.

The availability of reliable information is essential for the decision-maker, the manager and the scientist. Important parameters will condition the perception of forest genetic diversity and the degree of attention given to the subject, such as the type of resource that will be used, for which purpose, by which customer, in which region and over which period. While several initiatives have sought to better define FGR status and trends, a global assessment will bring added value to scattered efforts and help place genetic diversity issues in the local, national or global context in which they are best analysed and understood. A global perspective is increasingly necessary in some areas; industrial wood and wood products, and reproductive materials have become global commodities. However, a number of major challenges must be faced in designing such an assessment. In addition to biological and ecological features, the multiplicity of values and uses of forests, trees, biodiversity and genetic resources need to be considered with relevant stakeholders.

With the assistance of national partners and international collaborators, FAO will further elaborate on the foundation and process of a first global assessment of FGR.

4.7 REFERENCES

Convention on Biological Diversity. 1997a. *Recommendations for a core set of indicators of biological diversity.* UNEP/CBD/SBSTTA/3/Inf.13.

Convention on Biological Diversity. 1997b. *Indicators of forest biodiversity.* Working document. Prepared for the Meeting of the Liaison Group on Forest Biological Diversity. UNEP/CBD/SBSTTA/3/Inf.23.

Convention on Biological Diversity. 2004. *Indicators for assessing progress toward, and communicating, the 2010 target at the global level.* UNEP/CBD/SBSTTA/10/9.

Convention on Biological Diversity. 2004 *Indicators for assessing progress towards the 2010 target: trends in genetic diversity of domesticated animals, cultivated plants, and fish species of major socio-economic importance.* UNEP/CBD/SBSTTA/10/INF/14.

FAO. 1997. *State of the world's plant genetic resources for food and agriculture.* Rome.

FAO. 2001. *Global forest resources assessment 2000* (FRA 2000). FAO Forestry Paper 140. Rome.

FAO. 2002a. *Criteria and indicators for assessing the sustainability of forest management: conservation of biological diversity and genetic variation.* Forest Genetic Resources Working Paper FGR/37E, Forest Resources Development Service, Forest Resources Division. Rome.

FAO. 2002b. *Status and trends in indicators of forest genetic diversity.* Forest Genetic Resources Working Paper FGR/38E, Forest Resources Division. Rome.

Heywood, V.H. 1995. *Global biodiversity assessment.* Cambridge, England, UNEP/Cambridge University Press.

Holmgren, P. 2002. *Generic scope of global forest resources assessments - what are they about?* Background Paper 7.3. Presented at Kotka IV, Global Forest Resources Assessments - linking national and international efforts. 1-5 July 2002. Kotka, Finland.

Namkoong, G. 1986. Genetics and the future of the Forests. *Unasylva*, 38: 152. 1986/2: 2-18.

UNEP/CBD. 2001. *Global biodiversity outlook.* Montreal, Canada, Secretariat of the Convention on Biological Diversity.

II. Use of cryopreservation and reproductive technologies for conservation of genetic resources

5. The potential of cryopreservation and reproductive technologies for animal genetic resources convervation strategies

Sipke Joost Hiemstra, Tette van der Lende and Henri Woelders

5.1 SUMMARY

Ex situ conservation of genetic material from livestock and fish through cryopreservation is an important strategy to conserve genetic diversity in these species. Conservation strategies benefit from advances in cryopreservation and reproductive technologies. Choice of type of genetic material to be preserved for different species depends highly on objectives, technical feasibility (e.g. collection, cyropreservation), costs and practical circumstances.

5.2 INTRODUCTION

Global diversity in domestic animals is considered to be under threat. A large number of domestic animal breeds are endangered worldwide, in a critical status or already extinct. Of the 6 379 domestic animal breed populations, 9 percent are in critical condition and 39 percent are endangered (FAO, 2000). There is worldwide consensus on the global decline in domestic animal diversity and the need to conserve genetic diversity. The vast majority of aquatic genetic resources are found in wild populations of fish, invertebrates and aquatic plants. Domestication of aquatic species has not proceeded to the same level as it has in crop and livestock sectors. According to FAO, there are more than a thousand common aquatic species harvested by humans in major fisheries and thousands more species harvested in small-scale fisheries. The number of species in aquaculture is growing and several important species rely on the collection of brood stock or seed from natural populations.

In farm animals, trends in within-breed diversity are as important as between-breed diversity to be able to cope with changing requirements and future demands in breeding and selection. A small effective population size in rare or endangered breeds requires monitoring of within-breed diversity and conservation programmes to maintain within-breed diversity. Several authors also emphasized the reduction in effective population sizes of widely used domestic animal breeds (e.g. Weigel, 2001). Although introgression of genes for specific traits or characteristics from local breeds to commercial breeds has been very limited so far, Notter (1999) suggested that, as in plants, useful genes may exist in lowly productive types and recommended systematic programmes for genetic resources conservation, evaluation and use.

There are several options to conserve genetic diversity. In general, *in situ* conservation or conservation by utilization is preferred as a mechanism to conserve breeds. A breed has to evolve and adapt to changing environments and efforts should be promoted to create a need for products or functions of the breed. Conservation without further development of the breed or without expected future use is not a desirable strategy. However, in addition to *in situ* conservation, methods or techniques to maintain live animals outside their production or natural environment (*ex situ* live) or through cryopreservation of germplasm (*ex situ*) are set up to preserve (germplasm of) rare breeds as well as the more widely used commercial breeds. Moreover, cryopreservation of germplasm is a very good *ex situ* strategy to conserve existing allelic diversity for future use.

There is a growing interest in *ex situ* conservation strategies, serving a variety of objectives (ERFP, 2003). In many countries, *ex situ* conservation represents an integral component of conservation strategies (Blackburn, 2004). Some strategies focus primarily on preservation of germplasm of rare breeds, but in general there is consensus that *ex situ* collections should be established for all breeds with the aim to capture as much allelic or genetic diversity in conservation programmes as possible. Whereas *in situ* conservation or use of animal genetic resources is not necessarily dependent on high-tech approaches or facilities, the efficiency and efficacy of *ex situ* conservation strategies will certainly benefit from advances in cryopreservation and reproductive technology. Since *ex situ* conservation activities are in general rather costly, debate is ongoing on priorities, different methodologies and future use and benefits of cryopreservation and reproductive technology.

This chapter focuses on *ex situ* conservation. An overview of the state of the art in cryopreservation and reproductive technology for farm animals and fish is followed by a discussion on the implications for *ex situ* conservation strategies. This chapter is restricted to the main agricultural species; with regard to aquatic species, it deals with fish only and focuses on aquaculture rather than fisheries.

5.3 STATE OF THE ART IN CRYOPRESERVATION TECHNOLOGY

5.3.1 Cryobiologic principles

Cryopreservation allows virtually indefinite storage of biological material without deterioration over a time scale of at least several thousands of years (Mazur, 1985), but probably much longer. Important progress in cryobiology was achieved in the second half of the previous century. Much progress resulted from empirical studies. In later years, progress was also strongly stimulated by the development of fundamental theoretical cryobiology.

In so-called "slow cooling" methods, the biological material is cooled at a range of cooling rates that are fast enough to prevent "slow cooling damage" but slow enough to allow sufficient dehydration of the cells to prevent intracellular ice formation (IIF) (Mazur, Leibo and Chu, 1972). The dehydrated cells in the "unfrozen fraction" that remains between the masses of ice will ultimately reach a stable glassy state, or "vitrify". In so-called "vitrification methods", the water content is lowered before cooling by adding high concentrations of cryoprotective agents (CPA). Thus, no ice is formed at all, and the entire sample will vitrify. This allows fast cooling rates without risk of IIF. The CPA concentration of vitrification solutions can be minimized by using very high cooling and thawing rates. By using extremely high cooling rates, vitrification is possible even in complete absence of CPAs (Isachenko *et al.*, 2004).

5.3.2 Semen

Semen of most livestock species can be frozen adequately. In addition, dedicated freezing media and equipment for collecting, packing, freezing and inseminating semen have been developed and are available commercially for a large number of bird and mammal livestock species. In the cattle artificial insemination (AI) industry, in which bulls are selected for "freezability" of their semen, the post thaw semen quality is quite good, featuring 50 to 70 percent motile spermatozoa. Pregnancy or calving rate is the same as that of fresh semen, provided that higher sperm dosages are used for frozen sperm. For other mammalian species, the percentage of post-thaw motile sperm or membrane-intact sperm is generally somewhat lower, but a fair post-thaw viability can be expected for most species. For many species, the fertility of frozen semen is found to be lower than that of fresh semen. This may depend on the site of semen deposition, the morphology of the female genital tract, and the ability to detect heat or ovulation. For instance in sheep, very poor results are obtained with cervical semen deposition when using frozen rather than fresh ram semen (Molinia, Evans and Maxwell, 1996). There may be considerable differences between breeds and between males in the "freezability" of the semen. As a consequence, frozen semen of some genetically interesting breeds or males may not be suitable as a gene bank resource, or can be used only with a poor efficiency.

As regards avian livestock species, semen-freezing techniques for fowl, turkey, goose, and duck render a fair post-thaw sperm survival of up to 60 percent live spermatozoa. Reasonable insemination results with frozen-thawed semen have been reported for the major avian livestock species (Blesbois and Labbé, 2003; for more references, see Hammerstedt and Graham, 1992). However, there is a striking variation between studies in the reported percentages of fertilized eggs, as listed in Hammerstedt and Graham (1992), ranging from 9 to 91 percent. Moreover, the number of spermatozoa that gives maximal fertilization levels in chickens is much higher for frozen-thawed semen compared with fresh semen (Wishart, 1985).

More than 200 fish species with external fertilization have been tested for sperm cryopreservation (Blesbois and Labbé, 2003). The present state of the art for many species of fish seems to be adequate for the purpose of gene banking.

The insemination ratios used may vary according to species and procedure between 104 and 107 spermatozoa per egg. Even in fish species like the African catfish, in which semen can only be obtained by testis destruction or death of the male, enough semen can be obtained from one single male to produce close to 106 larvae (Viveiros, So and Komen, 2000). Thus, for gene bank purposes, storage of only one single vial or straw would be sufficient to generate plenty of progeny of that male.

The freezing media vary widely between the classes (mammals, birds, fish), but also between species within a class. Most media feature a saline or saccharide bulk osmotic support, a suitable CPA at concentrations varying from 0.2-1.5 M, and various protective macromolecular additives, mostly milk and egg yolk components, or lipid components from vegetal origin (Bousseau *et al.*, 1998). Milk or egg yolk is often used in media for mammalian semen. In mammalian semen, the egg yolk and milk components protect the spermatozoa during cooling, freezing and thawing (Watson, 1976). These additives are generally not used in freezing media for avian and fish species, although in a few studies with fish semen, egg yolk was found to confer protection against cryodamage (e.g. Cabrita *et al.*, 1998).

Glycerol is widely used as a suitable CPA in mammalian, bird, and fish species. However, in poultry it is found that glycerol is contraceptive, i.e. the semen must be washed free of glycerol after thawing (Hammerstedt and Graham, 1992). The type of CPA used varies widely between species, and occasionally within one species, a CPA has been successfully used in one study and was found to be unsuitable in another study with the same species (Viveiros, So and Komen, 2000). Glycerol is used in most mammalian species. In avian species, dimethyl sulphoxide (DMSO), Ethylene glycol (EG), dimethylacetamide (DMA) and dimethylformamide (DMF) are also frequently used. In fish species, glycerol, DMA, DMF, DMSO and methanol are often used.

Semen is generally cryopreserved with "slow cooling" methods. Optimal cooling rates for freezing semen are mostly found between 10 and 100 °C/min. To some extent, the reported differences may be related to the use of different types of CPA and different CPA concentrations. An extreme example is that fowl semen can be effectively frozen at a cooling rate of approximately 600 °C/min when using dimethylacetamide as CPA, but not using glycerol (Woelders *et al.*, in preparation). CPAs may differ widely in the cell membrane permeability, and may also affect the membrane permeability to water. These parameters greatly affect the velocity of dehydration, and therewith the optimal range of cooling rates.

5.3.3 Oocytes

In the last ten years, considerable progress has been made with cryopreservation of oocytes. Viable oocytes have been recovered after freezing and thawing in a great number of species (see references in Woelders, Zuidberg and Hiemstra, 2003). Successes have been reported in post-thaw oocyte maturation, fertilization, and embryo development in a number of species. Live-born young from embryos produced from cryopreserved oocytes have been reported in cattle (Otoi *et al.*, 1996; Abe *et al.*, 2005), mice (Frydman *et al.*, 1997), (Stachecki *et al.*, 2002), rats (Nakagata, 1992), horses (Maclellan *et al.*, 2002) and humans (Stachecki and Cohen, 2004). The present efficiency and reliability of using frozen thawed oocytes for generating offspring is still much lower than with cryopreserved embryos. Freezing oocytes of avian and fish species is not successful (Blesbois and Labbé, 2003), however, largely because of the large size, the high lipid content, and the polar organization (vegetal and animal pole) of bird and fish ova.

5.3.4 Embryos or embryonic cells

In cattle, cryopreservation of embryos is highly successful. Both slow freezing and vitrification protocols are effective. The success of cryopreservation depends on the stage of the embryo, that is, especially good results are obtained with blastocysts. Cryopreservation of embryos resulting in live offspring has been reported for most of the important (mammalian) livestock species (reviewed by Paynter *et al.*, 1997; see also Cognie *et al.*, 2003; Squires *et al.*, 2003). Cryopreservation of pig embryos has long been problematic due to extreme chilling sensitivity and high lipid content of the pig embryos. However, recent studies have focused on overcoming these problems and produced successful vitrification methods for cryopreservation of pig embryos (e.g. Vajta, 2000; Nagashima *et al.*, 1999; Dobrinsky *et al.*, 2000).

Embryo cryopreservation is not viable in birds (Blesbois and Labbé, 2003) and fish (Hagedorn *et al.*, 2004) species, largely because of the same limitations as in the case of avian and fish oocytes, i.e. the large size, the high lipid content, and the polar organization of the ova and the early embryos of fish and birds. However, in birds and fish species, cryopreservation of isolated embryonic cells is an option. Post-thaw survival of blastomeres was demonstrated in rainbow trout, carp and medaka (see references in Blesbois and Labbé, 2003). Embryonic cells and recipient embryos can be used to produce chimeric embryos. Provided that the gonads become populated with primordial germ cells from the donor embryo, such chimeric embryos can be used to produce future progeny of the donor genotype (e.g. Nakagawa, Kobayashi and Ueno, 2002). In chicken, the primordial germ cells (PGC) can be specifically harvested. Recently, improvement of the efficiency of producing chimerae with donor genotype germinal cells was achieved by depleting PGC from the recipient embryos using busulfan (Song *et al.*, 2005).

5.3.5 Somatic cells

Cryopreservation of somatic cells proved to be possible for a number of cell types. In early studies, the methods came down to adding 5 to 10 percent of a suitable cryoprotectant, such as glycerol or DMSO to the suspension of cells in culture medium, and placing tubes with a few millilitres of that suspension at -80 °C in a mechanical freezer. In fact, this simple procedure is still effectively used today. Obviously, with this simple procedure the rate of cooling cannot be controlled; in fact in many publications the cooling rate is unknown. There are only a few studies in which controlled rate freezers were used, e.g. with skin fibroblast.

5.3.6 Further progress

More attention to fundamental aspects of cryobiology should enable further progress in cryopreservation methods. A fundamental approach has been taken in many studies concerning mammalian semen and embryos (Gilmore *et al.*, 1996, Liu *et al.*, 2000; Chaveiro *et al.*, 2004), but fewer concerning avian and fish semen (e.g. Viveiros *et al.*, 2001). Recently, a theoretical model was presented to predict the optimal cooling programme for "slow cooling" freezing methods (Woelders and Chaveiro, 2004). The model indicated that a non-linear cooling profile could give better results than linear freezing programmes. This and other models (cf. Liu *et al.*, 2000) also demonstrate that the optimal cooling rate can be expected to be inversely related to the CPA concentration, and in fact, this is found in empirical studies. Therefore, it is important to address both factors in empirical optimization studies. It can also mean that a lower concentration of CPA would become feasible provided that a higher cooling rate is used. Further improvement could result from preventing delayed ice formation or "supercooling", e.g. by using so-called "directional solidification" methods (Woelders *et al.*, 2005). Improving the freezing methods can raise the general level so that even the semen of "bad freezers" would have an adequate post-thaw sperm survival (ibid.).

Attempts to vitrify spermatozoa have not been successful to date. It has recently been shown that vitrification of human spermatozoa is possible in the absence of CPA by using an extremely high cooling rate of 720 000 °C/min (Isachenko *et al.*, 2004). In this way, damage due to the presence of the CPA, chilling injury and ice formation may be avoided. Further improvement of vitrification techniques is especially important for freezing cells that are sensitive to chilling, e.g. to "outrun" spindle microtubule depolymerization in metaphase II oocytes. Very high cooling rates can be applied in the open-pulled straw (OPS) technique (Vajta *et al.*, 1998) or by using the cryoloop (Lane, Schoolcraft and Gardner, 1999; Isachenko *et al.*, 2004). However, interrupted slow cooling methods can also be highly effective, as a fully normal and functional spindle can reform after thawing (Stachecki and Cohen, 2004).

5.4 STATE OF THE ART IN REPRODUCTIVE TECHNOLOGY

5.4.1 Artificial insemination (AI)

In several species, AI techniques and strategies have been improved and knowledge on the fate of sperm in the female genital tract (e.g. phagocytosis) improved during the last decades. However, there are large differences between species in insemination techniques and pregnancy rates using fresh or frozen semen. In cattle and pigs, existing AI infrastructure allows easy collection and future use of semen, but only in cattle has the use of frozen semen replaced the use of fresh semen. In pig production, disadvantages of using frozen semen (reduced fertility, high freezing, storage and transport costs) are still greater than the advantages.

In sheep, surgical (laparoscopic) AI gives much better pregnancy rates than cervical AI. However, laparoscopic AI is more laborious and also more invasive than cervical AI. Molinia *et al.* (1996) showed that the difference in pregnancy rates between surgical and non-surgical AI was even larger with frozen semen than with fresh semen: 70 percent versus 20 percent pregnancy with non-surgical AI (with 180 x 106 sperm) vs. surgical AI (with 10 x 106 sperm). It is believed that frozen-thawed sperm are less motile and lack stamina to transverse the highly viscous cervical mucus, but phagocytosis of the sperm by leukocytes is also considered a cause of the reduced fertility. Development of a non-surgical technique to reach the oviductal end of the uterine horns as closely as possible would enhance the efficiency and ease of use of cryopreserved semen in sheep. Such deep intrauterine insemination techniques have been developed in pigs (reviewed by Vazguez *et al.*, 2005) and may in general contribute to the more efficient use of semen (less sperm per insemination).

AI can be used successfully in poultry, but is not used extensively in any domestic avian species except turkeys where it is used almost exclusively for commercial flock production (reviewed by Donoghue and Wishart, 2000).

5.4.2 Embryo transfer

Surgical embryo transfer could be possible in principle in all mammalian livestock species. In contrast, non-surgical embryo transfer is only possible in cattle (routinely performed), horses and pigs, although still not as efficient as in cattle and horses. For embryo transfer purposes, embryos can either be flushed from donors or can be produced *in vitro*. Surgical embryo collection is in principle possible in all mammalian livestock species. In contrast, non-surgical embryo collection is only possible in cattle and horses. After surgically shortening the long uterine horns of the pig, non-surgical recovery of embryos has also been proven possible in pigs (Hazeleger *et al.*, 1989). Although ethical issues have prevented the further use of this method, it may be used in specific situations, e.g. to collect large numbers of embryos in very rare pig breeds in a relatively short time. The efficiency of non-surgical embryo collection in cattle, and to a lesser extent in horses, can be improved by hormonal induction of superovulation.

In vitro production of embryos by *in vitro* maturation and fertilization of oocytes is possible for major livestock breeds but the efficiency varies between species. Oocytes for this purpose can either be collected by aspiration of immature oocytes from ovaries from slaughtered (or deceased) animals or by the use of ovum pick-up techniques in live animals. The latter techniques are presently mainly in use in cattle and horses but could also be used in other livestock species.

5.4.3 Reproductive cloning

Reproductive cloning involves the collection of oocytes, culture and *in vitro* maturation of oocytes, enucleation of oocytes, transfer of (somatic) nuclei to, or fusion of the somatic cells with, enucleated oocytes, culture of the resulting embryos, and finally, embryo transfer into recipients of the same or a highly related species (reviewed by Galli, Lagutina and Lazzari, 2003). The use of nuclear transfer means that the original mitochondrial genotype of the nucleus donor is lost.

In mammals, live offspring have been obtained from embryos generated from somatic cells in a number of species, i.e. sheep, cattle, mice, pigs, goats, horses, rabbits and cats. Until now, cloning has failed in rats, rhesus monkeys and dogs. Remarkably, some success (embryo development but no live offspring) was even obtained when bovine oocytes were used as recipients for somatic nuclei from other mammalian species (e.g. Dominko *et al.*, 1999). It must be emphasized, however, that current techniques are inadequate to be used safely and efficiently for procreation. In all published research, only a small proportion of embryos produced by using somatic cells developed into live offspring, i.e. typically less than 4 percent (Paterson *et al.*, 2003). The low overall success rate is the cumulative result of inefficiencies at each stage of the cloning process. Many pregnancies are terminated by abortion and full-term pregnancies frequently result in abnormal offspring. It therefore seems that current cloning techniques introduce errors that affect prenatal development. Even apparently healthy live-born offspring could have anomalies that only become apparent later in life, or in the next generation of animals. On a long time horizon it is very likely that cloning methodology will become both reliable and efficient.

In fish, successful cloning has been reported by Lee *et al.* (2002). In their experiments with zebrafish, an overall success rate of 2 percent was achieved. To the best knowledge available, no successful cloning has been reported in poultry.

5.4.4 Miscellaneous emerging reproductive technologies

Transplantation of ovarian tissue and germ cells (e.g. PGC or spermatogonial stem cells [SSC]) are emerging technologies with potential for future use in conservation programmes. Autotransplantation of ovarian tissue has been developed to restore fertility in women after aggressive chemotherapy resulting in ovarian failure. Successful transplantation of ovarian tissue has been reported in rodents, sheep, marmoset monkeys and humans (references in Donnez *et al.*, 2004). Successful whole sheep ovary cryopreservation and autotransplantation has recently been reported (Revel *et al.*, 2004). The potential of ovary transplantation as a tool in genetic conservation is underlined by the work of Dorsch *et al.* (2004). Their experiments demonstrate that transplantation of rat ovaries can be used as a tool for the rescue of rat strains where females are unable to reproduce despite having normal ovarian cycles.

Germ cell transplantation research has been developed as a unique approach for the study of gametogenesis and germ line manipulation. To date, successful germ cell transplantations have been reported in several livestock species, e.g. transplantation of SSC in cattle (Izadyar *et al.*, 2003) and goats (Honaramooz *et al.*, 2003), but also in poultry (e.g. Park *et al.*, 2003) and fish (e.g. Takeuchi, Yoshizaki and Takeuchi, 2003). As far as application of this technology in fish is concerned, fascinating results of allogenic transplantation in rainbow trout have been reported (ibid.). By transplanting PGC of donors into the peritoneal cavity of hatching recipient embryos, live fry with donor-derived phenotype were produced from gametes of PGC-recipients.

Although many hurdles have to be overcome, these emerging technologies may enable production of gametes or offspring of rare or extinct breeds in the long term by abundantly available individuals of related common breeds. The first steps to overcome limitations for homologous transfer of ovarian tissue and germ cells are now underway; the development of an effective recipient preparation protocol in mice (Brinster *et al.*, 2003) is an example.

5.5 IMPLICATIONS FOR *EX SITU* CONSERVATION STRATEGIES

The choice of type of material to be preserved and sampling strategies depends, *inter alia*, on the objectives of cryopreservation programmes (ERFP, 2003). Decisions will be different between species because of variation in technical feasibility, costs and practical circumstances for cryopreservation of different types of material. In general, cryopreservation and associated reproductive technologies are costly; the main limitations for extensive development of *ex situ* collections are high costs of collection and limited use of preserved material (FAO, 2004). Costs of sampling, collection, freezing, storage and use of genetic material differ between species, and optimum strategies depend on local circumstances, availability of technology and costs of labour and facilities. Gandini and Pizzi (2003) reviewed the literature on conservation costs (*in situ* and *ex situ*) and concluded that published information on *ex situ* conservation costs was very limited and not very timely. Labroue *et al.* (2001) calculated total costs for creating pig semen storage among four European countries at about 30 000 euros per breed and 15 euro per dose. Costs of cryopreservation of pig semen in the Netherlands (1999-2001) were estimated at approximately 10 euro per dose, based on costs for labour, laboratory materials and infrastructure, assuming a freezing capacity of six ejaculates per day. Transport costs of the semen from the AI centre to the freezing facility and costs for semen collection are not included in this figure.

The Centre for Genetic Resources (CGN) observed that costs of collecting and freezing of semen of different species vary from less than 1 euro per dose (cattle) to more than 20 euro per dose (sheep and poultry). The higher costs in sheep and poultry are due to the much higher handling, training, collection and freezing costs per dose of semen and the lack of AI infrastructure in these species.

As an alternative for semen collection and freezing of ejaculated semen, CGN concluded that collection and freezing of epididymal semen of culled rams is a cost-effective method to conserve genetic diversity in sheep breeds (Woelders *et al.*, in preparation).

Differences in generation interval and reproductive rates between species may also influence decision-making in conservation programmes. In some species it is possible to regenerate a breed very quickly with inexpensive, sometimes less sophisticated methods, compared to other species. For example, in fish the regeneration of an extinct species with stored semen is feasible through backcrossing since fish species have a low generation interval and a high annual turnover. In contrast, such a strategy in horse or cattle would be time-consuming and extremely expensive. In these species, cryo-banking of embryos rather than sperm is highly preferable.

Costs of embryo collection and freezing are much higher than those for semen collection and freezing. However, regeneration costs using embryos are much lower than those for semen (repeated backcrossing). Many conservation programmes focus on freezing of semen only. If the aim is to conserve breeds and taking into account the loss of mitochondrial DNA and the time lag to re-establish a breed by backcrossing, collection and cryopreservation of embryos is underrated. In this context, cryopreservation of somatic cells does not seem to be a good alternative for cryopreservation of embryos, even if the efficiency of cloning is largely improved. Upfront costs of freezing somatic cells may be very low, but mitochondrial DNA is not conserved and the efficiency of subsequent steps in reproductive cloning can never beat the efficiency of cryo-conservation and implantation of embryos. Storage of both oocytes and semen may also be efficient in terms of sampling and freezing costs. However, high costs are associated with *in vitro* fertilization and ovum pick-up (OPU). Costs are expected not to be lower overall than when using embryos instead of semen plus oocytes.

When survival of material after freezing/thawing improves and the chance of pregnancy increases, costs of sampling and freezing of gene bank material will drop because less genetic material is needed in the gene bank to generate a sufficient number of live offspring. Furthermore, if freezability of semen of genetically important males can be improved substantially (especially in the case of "bad freezers"), sampling costs will drop even more.

Cryopreservation technology strongly affected reproduction in livestock. Ironically, unsustainable use of cryopreservation can cause a decline in genetic diversity, but at the same time its use is beneficial when applied to conservation programmes. For example, in dairy cattle the combined use of genetic evaluation, AI and frozen semen, and more recently, several other reproductive technologies, has resulted in high genetic gain in the Holstein Friesian (HF) breed and thus stimulated their worldwide, large-scale use at the cost of local dairy breeds and a decline in the effective population size of the HF breed. As a side effect, however, know-how on cryopreservation, AI and other reproductive technologies developed for use in cattle has turned out to be of major importance for the conservation of breeds of various species.

5.6 CONCLUSIONS

- In *ex situ* conservation programmes with the aim to conserve breeds, collection and cryopreservation of embryos is underrated, and should be given more attention.
- Advances in cryopreservation and reproductive technology have contributed and will continue to contribute to the efficiency and effectiveness of conservation programmes.
- Species in which cryopreservation and reproductive technology are less developed can benefit from advances in other species.
- Decision-makers in conservation programmes should regularly reconsider the balance between objectives, costs, technical feasibility and practical feasibility.

5.7 REFERENCES

Abe, Y., Hara, K., Matsumoto, H., Kobayashi, J., Sasada, H., Ekwall, H., Rodriguez-Martinez, H. & Sato, E. 2005. Feasibility of a nylon mesh holder for vitrification of bovine germinal vesicle oocytes in subsequent production of viable blastocysts. *Biology of Reproduction.* (in press)

Blackburn, H.D. 2004. Development of national animal genetic resource programs. *Reproduction, Fertility and Development*, 16: 27-32.

Blesbois, E. & Labbé, C. 2003. Main improvements in semen and embryo cryopreservation for fish and fowl. *In* D. Planchenault, ed. *Workshop on Cryopreservation of Animal Genetic Resources in Europe*, Paris. pp. 55-65. ISBN 2-908447-25-8.

Bousseau, S., Brillard, J.P., Marguant-Le Guienne, B., Guerin, B., Camus, A. & Lechat, M. 1998. Comparison of bacteriological qualities of various egg yolk sources and the *in vitro* and *in vivo* fertilizing potential of bovine semen frozen in egg yolk or lecithin-based diluents. *Theriogenology*, 50: 699-706.

Brinster, C.J., Ryu, B.Y., Avarbock, M.R., Karagenc, L., Brinster, R.L. & Orwig, K.E. 2003. Restoration of fertility by germ cell transplantation requires effective recipient preparation. *Biology of Reproduction*, 69: 412-420.

Cabrita, E., Alvarez, R., Anel, L., Rana, K.J. & Herraez, M.P. 1998. Sublethal damage during cryopreservation of rainbow trout sperm. *Cryobiology*, 37: 245-253.

Chaveiro, A., Liu, J., Mullen, S., Woelders, H. & Critser, J.K. 2004. Determination of bull sperm membrane permeability to water and cryoprotectants using a concentration-dependent self-quenching fluorophore. *Cryobiology*, 48: 72-80.

Cognie, Y., Baril, G., Poulin, N. & Mermillod, P. 2003. Current status of embryo technologies in sheep and goat. *Theriogenology*, 59: 171-188.

Dobrinsky, J.R., Pursel, V.G., Long, C.R. & Johnson, L.A. 2000. Birth of piglets after transfer of embryos, cryopreserved by cytoskeletal stabilization and vitrification. *Biology of Reproduction*, 62: 564-570.

Dominko, T., Mitalipova, M., Haley, B., Beyhan, Z., Memili, E., McKusick, B. & First, N.L. 1999. Bovine oocyte cytoplasm supports development of embryos produced by nuclear transfer of somatic cell nuclei from various mammalian species. *Biology of Reproduction*, 60: 1496-1502.

Donnez, J., Dolmans, M.M., Demylle, D., Jadoul, P., Pirard, C., Squiffket, J., Martinez-Madrid, B., & Van Langendockt, A. 2004. Livebirth after orthotopic transplantation of cryopreserved ovarian tissue. *The Lancet*, 364: 1405-1410.

Donoghue, A.M. & Wishart, G.J. 2000. Storage of poultry semen. *Animal Reproduction Science*, 62: 213-232.

Dorsch, M., Wedekind, D., Kamino, K. & Hedrich, H.J. 2004. Orthotopic transplantation of rat ovaries as a tool for strain rescue. *Laboratory Animals*, 38: 307-312.

ERFP. 2003. Guidelines for the constitution of national cryopreservation programmes for farm animals. *In* S.J. Hiemstra, ed. *European Regional Focal Point on Animal Genetic Resources*, Publication No. 1.

FAO. 2000. *World watch list for domestic animal diversity*, 3rd edition. Rome.

FAO. 2004. *Conservation strategies for animal genetic resources*, by D.R. Notter. Background study paper no. 22. Commission on Genetic Resources for Food and Agriculture. Rome.

Frydman, N., Selva, J., Bergere, M., Auroux, M. & Maro, B. 1997. Cryopreserved immature mouse oocytes: a chromosomal and spindle study. *Journal of Assisted Reproduction and Genetics*, 14: 617-623.

Galli, C., Lagutina, I. & Lazzari, G. 2003. Introduction to cloning by nuclear transplantation. *Cloning and Stem Cells*, 5: 223-232.

Gandini G. & Pizzi, F. 2003. *In situ* and *ex situ* conservation techniques: financial aspects. In *Proceedings of the Workshop on Cryopreservation of Animal Genetic Resources in Europe*, Paris, 23 February, 2003.

Gilmore, J.A., Du, J., Tao, J., Peter, A.T. & Critser, J.K. 1996. Osmotic properties of boar spermatozoa and their relevance to cryopreservation. *Journal of Reproduction and Fertility*, 107: 87-95.

Hagedorn, M., Peterson, A., Mazur, P. & Kleinhans, F.W. 2004. High ice nucleation temperature of zebrafish embryos: slow-freezing is not an option. *Cryobiology*, 49: 181-189.

Hammerstedt, R.H. & Graham J.K. 1992. Cryopreservation of poultry sperm: the enigma of glycerol. *Cryobiology*, 29: 26-38.

Hazeleger, W., Van der Meulen, J. & Van der Lende, T. 1989. A method for transcervical embryo collection in the pig. *Theriogenology*, 32: 727-734.

Honaramooz, A., Behboodi, E., Blash, S. & Megee, S.O. 2003. Germ cell transplantation in goats. *Molecular Reproduction and Development*, 64: 422-428.

Isachenko, V., Isachenko, E., Katkov, I.I., Montag, M., Dessole, S., Nawroth, F. & Van Der Ven, H. 2004. Cryoprotectant-free cryopreservation of human spermatozoa by vitrification and freezing in vapor: effect on motility, DNA integrity, and fertilization ability. *Biology of Reproduction*, 71: 1167-1173.

Izadyar, F., Den Ouden, K., Stout, T.A.E., Stout, J., Coret, J., Lankveld, D.P.K., Spoormakers, T.J.P., Colenbrander, B., Oldenbroek, J.K., Van der Ploeg, K.D., Woelders, H., Kal, H.B. & De Rooij, D.G. 2003. Autologous and homologous transplantation of bovine spermatogonial stem cells. *Reproduction*, 126: 765-774.

Labroue, F., Luquet, M., Guillouet, P., Bussiere, J.F., Glodek, P., Wemheuer, W., Gandini, G., Pizzi, F., Delgado J.V., Poto, A., Peinado, B., Sereno, R.B. & Ollivier, L. 2001. Pig semen banks in Europe. *In* L. Ollivier, F. Labroue, P. Glodek, G. Gandini & J.V. Delgado, eds. *Pig genetic resources in Europe.* Wageningen, The Netherlands, Wageningen Pers.

Lane, M., Schoolcraft, W.B. & Gardner, D.K. 1999. Vitrification of mouse and human blastocysts using a novel cryoloop container-less technique. *Fertility and Sterility*, 72:1073-1078.

Lee, K.-Y., Huang, H., Ju, B., Yang, Z. & Lin, S. 2002. Cloned zebrafish by nuclear transfer from long-term-cultured cells. *Nature Biotechnology*, 20: 795-799.

Liu, J., Woods, E.J., Agca, Y., Critser, E.S. & Critser, J.K. 2000. Cryobiology of rat embryos II: A theoretical model for the development of interrupted slow freezing procedures. *Biology of Reproduction*, 63: 1303-1312.

Maclellan, L.J., Carnevale, E.M., Coutinho da Silva, M.A., Scoggin, C.F., Bruemmer, J.E. & Squires, E.L. 2002. Pregnancies from vitrified equine oocytes collected from super-stimulated and non-stimulated mares. *Theriogenology*, 58: 911-919.

Mazur, P. 1985. Basic concepts in freezing cells. *In* L.A. Johnson & K. Larsson, eds. *Proceedings of the First International Conference on Deep Freezing of Boar Semen*, Uppsala, Sweden, pp. 91-111.

Mazur, P., Leibo, S.P. & Chu, E.H. 1972. A two-factor hypothesis of freezing injury. Evidence from Chinese hamster tissue-culture cells. *Experimental Cell Research*, 71: 345-355.

Molinia, F.C., Evans, G. & Maxwell, W.M.C. 1996. Fertility of ram spermatozoa pellet-frozen in zwitterion-buffered diluents. *Reproduction, Nutrition, Development*, 36: 21-29.

Nagashima, H., Cameron, R.D.A., Kuwayama, M., Young, M., Beebe, L., Blackshaw, A.W. & Nottle, M.B. 1999. Survival of porcine delipated oocytes and embryos after cryopreservation by freezing or vitrification. *Journal of Reproduction and Development*, 45: 167-176.

Nakagata, N. 1992. Cryopreservation of unfertilized rat oocytes by ultrarapid freezing. *Jikken Dobutsu*, 41: 443-447 (Japanese).

Nakagawa, M., Kobayashi, T. & Ueno, K. 2002. Production of germline chimera in loach *(Misgurnus anguillicaudatus)* and proposal of new method for preservation of endangered fish species. *Journal of Experimental Zoology*, 293: 624-631.

Notter, D.R. 1999. The importance of genetic diversity in livestock populations of the future. *Journal of Animal Science*, 77: 61-69.

Otoi, T., Yamamoto, K., Koyama, N., Tachikawa, S. & Suzuki, T. 1996. A frozen-thawed *in vitro*-matured bovine oocyte derived calf with normal growth and fertility. *Journal of Veterinary and Medical Science*, 58: 811-813.

Park, T.S., Hong, Y.H., Kwon, S.C., Lim, J.M. & Han, J.Y. 2003. Birth of germline chimeras by transfer of chicken embryonic germ (EG) cells into recipient embryos. *Molecular Reproduction and Development*, 65: 389-395.

Paterson, L., DeSousa, P., Ritchie, W., King, T. & Wilmut, I. 2003. Application of reproductive biotechnology in animals: implications and potentials. Applications of reproductive cloning. *Animal Reproduction Science*, 79: 137-143.

Paynter, S., Cooper, A., Thomas, N. & Fuller, B. 1997. Cryopreservation of multicellular embryos and reproductive tissues. *In* A.M. Karow & J.K. Critser, eds. *Reproductive tissue banking; scientific principles*. Academic Press. pp. 359-397.

Reve, H. Gustafsson, T. Katila, H. Kindahl & E. Ropstad, eds. Proceedings of the 14th International Congress on Animal Reproduction, **2000.** *Animal Reproduction Science*, 60-61: 357-364.

Revel, A., Elami, A., Bor, A., Yavin, S., Natan, Y. & Arav, A. 2004. Whole sheep ovary cryopreservation and transplantation. *Fertility and Sterility*, 82: 1714-1715.

Song, Y., D'Costa, S., Pardue, S.L. & Petitte, J.N. 2005. Production of germline chimeric chickens following the administration of a busulfan emulsion. *Molecular Reproduction and Development*, 70(4): 438-444.

Squires, E.L., Carnevale, E.M., McCue, P.M. & Bruemmer, J.E. 2003. Embryo technologies in the horse. *Theriogenology*, 59: 151-170.

Stachecki, J.J. & Cohen, J. 2004. An overview of oocyte cryopreservation. *Reprod. Biomed Online*, 9:152-163.

Stachecki, J.J., Cohen, J., Schimmel, T. & Willadsen, S.M. 2002. Fetal development of mouse oocytes and zygotes cryopreserved in a nonconventional freezing medium. *Cryobiology*, 44: 5-13.

Takeuchi, Y., Yoshizaki, G. & Takeuchi T. 2003. Generation of live fry from intraperitoneally transplanted primordial germ cells in rainbow trout. *Biology of Reproduction*, 69: 1142-1149.

Vajta, G. 2000. Vitrification of the oocytes and embryos of domestic animals. In Animal reproduction: research and practice II. *In* M. Forsberg, T. Greve.

Vajta, G., Holm, P., Kuwayama, M., Booth, P.J., Jacobsen, H., Greve, T. & Callesen, H. 1998. Open pulled straw (OPS) vitrification: a new way to reduce cryoinjuries of bovine ova and embryos. *Molecular Reproduction and Development*, 51: 53-58.

Vazguez, J.M., Martinez, E.A., Roca, J., Gil, M.A., Parrilla, I., Cuello, C., Carvajal, G., Lucas, X. & Vazguez, J.L. 2005. Improving the efficiency of sperm technologies in pigs: the value of deep intrauterine insemination. *Theriogenology*, 63: 536-547.

Viveiros, A.T., So, N. & Komen, J. 2000. Sperm cryopreservation of African catfish, *Clarias gariepinus*: cryoprotectants, freezing rates and sperm: egg dilution ratio.

Theriogenology, 54: 1395-1408.

Viveiros, A.T., Lock, E.J., Woelders, H. & Komen, J. 2001. Influence of cooling rates and plunging temperatures in an interrupted slow-freezing procedure for semen of the African catfish, *Clarias gariepinus. Cryobiology*, 43: 276-287.

Watson, P.F. 1976. The protection of ram and bull spermatozoa by the low density lipoprotein fraction of egg yolk during storage at 5 °C and deep freezing. *Journal of Thermal Biology*, 1: 137-141.

Weigel, K.A. 2001. Controlling inbreeding in modern breeding programs. *Journal of Dairy Science*, 84 (Suppl. E), E177-E184.

Wishart, G.J. 1985. Quantitation of the fertilizing ability of fresh compared with frozen and thawed fowl spermatozoa. *British Poultry Science*, 26: 375-380.

Woelders, H. & Chaveiro, A. 2004. Theoretical prediction of 'optimal' freezing programmes. *Cryobiology*, 49: 258-271.

Woelders, H., Matthijs, A., Zuidberg, C.A. & Chaveiro, A.E. 2005. Cryopreservation of boar semen: equilibrium freezing in the cryomicroscope and in straws. *Theriogenology*, 63: 383-395.

Woelders, H., Zuidberg, C.A. & Hiemstra, S.J. 2003. Applications, limitations, possible improvements and future of cryopreservation for livestock species. In *Proceedings of the Workshop on Cryopreservation of Animal Genetic Resources in Europe.* Paris, 23 February 2003. pp. 67-76.

6. Status of cryopreservation technologies in plants (crops and forest trees)

Bart Panis and Maurizio Lambardi

6.1 SUMMARY

Over the past decades, plant cryopreservation technologies have been evolving rapidly, opening the door to the possibility of long-term storage of valuable genetic resources of many crop and forest species. From the original slow-cooling approach, research has moved to easier and more reproducible techniques, which allow the complete vitrification of extra- and intra-cellular liquids. This chapter describes concisely the procedures that have been proposed in time for the cryopreservation of a wide range of tissues and organs, such as cell suspensions, embryogenic callus, pollen, meristematic tissues, seeds and embryo axes. In addition, the most important achievements in the cryopreservation of herbaceous, hardwood and softwood species are discussed.

6.2 INTRODUCTION

It is estimated that up to 100 000 plants, representing more than one-third of the world's plant species, are currently threatened or face extinction in the wild (BGCI, 2005). Preservation of the plant biodiversity is essential for classical and modern (genetic engineering) plant breeding programmes. Moreover, this biodiversity provides a source of compounds to the pharmaceutical, food and crop protection industries. Since the 1970s, large numbers of landraces and wild relatives of cultivated crops have been sampled and stored in *ex situ* gene banks. It is estimated that six million samples of plant genetic resources are held in national, regional, international and private gene bank collections around the world (IPGRI, 2004). Storage of desiccated seeds at low temperature, the most convenient method to preserve plant germplasm, is not applicable to crops that do not produce seed, such as bananas, nor to those with recalcitrant seed - non-orthodox seed that cannot be dried to moisture contents that are low enough for storage, for instance, many tropical trees, nor to plant species that are propagated vegetatively to preserve the unique genomic constitution of cultivars, such as fruit and several timber and ornamental trees. For these species, although clonal orchards play a pre-eminent role in assisting conservation programmes, their maintenance requires large areas of land and high running costs, mainly for pruning operations, weed and pest management and irrigation. Further, they are prone to environmental stresses such as heavy frosts and flooding and to hazards such as pests, diseases and genetic alterations; hence, valuable germplasm can be easily lost (genetic erosion). *In vitro* collections, established for some vegetatively propagated species and maintained by means of traditional micropropagation, is labour-intensive as well, and there is always the risk of losing accessions due to contamination, human error or somaclonal variation (i.e. mutations that occur spontaneously in tissue culture, with a frequency that increases with repeated subculturing).

Today, tissue culture technology offers two major options to *ex situ* conservation of valuable plant germplasm - slow growth storage (i.e. the medium-term conservation of stock cultures at few degrees above zero) and cryopreservation (Benson, 1999). Cryopreservation, or freeze-preservation at ultra-low temperature (-196 °C, i.e., the temperature of liquid nitrogen), is a sound alternative for the long-term conservation of plant genetic resources since biochemical and most physical processes are completely arrested under these conditions. As such, plant material can be stored for unlimited periods. Moreover, in addition to its use for the conservation of genetic resources, cryopreservation proved to be extremely useful for the safe long-term storage of plant tissues with specific characteristics, such as medicinal- and alkaloid-producing cell lines, hairy root cultures, and genetically transformed (Elleuch *et al.*, 1998) and transformation-competent culture lines (Gordon-Kamm *et al.*, 1990). Recently, it was also proven that cryotherapy can be successfully applied to eradicate viruses from plum, banana and grape (Brison *et al.*, 1997; Helliot *et al.*, 2002; Wang *et al.*, 2003). However, despite the fact that cryogenic procedures are now being developed for an increasing number of recalcitrant seeds and *in vitro* tissues/organs, the routine utilization of cryopreservation for the preservation of plant biodiversity is still limited.

6.3 THE THEORETICAL BASIS OF PLANT CRYOPRESERVATION

Cryopreservation of biological tissues can be successful only if intra-cellular ice crystal formation is avoided since this causes irreversible damage to cell membranes, thus destroying their semi-permeability. In nature, some plant species adopted systems where ice crystal formation at sub-zero temperatures can be avoided through the synthesis of specific substances (such as sugars, proline and proteins) that lower the freezing-point in the living plant cells, resulting in "supercooling". Such "avoidance" of crystallization, while still maintaining a minimal moisture level needed to maintain viability, is not possible when dealing with ultra-low temperatures of cryopreservation. Crystal formation, without an extreme reduction of cellular water, can only be prevented through vitrification. Vitrification refers to the physical transition process from an aqueous solution to an amorphous and glassy (i.e. non-crystalline) (Sakai, 2000). Two requirements must be met for a cell to vitrify: rapid freezing rates and a concentrated cellular solution. Rapid freezing rates (6 °C/sec) are normally obtained by plunging explants enclosed in a cryovial into liquid nitrogen. Higher cooling rates can be obtained by enclosing the meristems in semen straws, resulting in cooling rates of about 60 °C/sec, or using a "droplet freezing protocol" where the material is placed on aluminium foil strips plunged directly into liquid nitrogen, giving rise to cooling rates of 130 °C/sec (Panis, Piette and Swennen, 2005; Schäfer-Menuhr, Schumacher and Mix-Wagner, 1997).

The cell cytosol can be concentrated through air drying, freeze dehydration, application of penetrating or non-penetrating substances (cryoprotectants), or adaptive metabolism (hardening). For a solution to be vitrified at high cooling rates, a reduction in water content to at least 20 to 30 percent is required. For dehydration, the following techniques are applied:

6.3.1 Air drying

Samples are usually dried by the sterile airflow of a laminar airflow cabinet. However, with this method there is no control of temperature and air humidity, both strongly influencing evaporation rates. More reproducible is the air-drying method that uses closed vials containing a fixed amount of silica gel (Uragami, Sakai and Nagai, 1990).

6.3.2 Freeze dehydration

Since plant cells rarely contain ice-nucleating agents, during a slow cooling process, crystallization is initiated first in the extra-cellular spaces. Since only a proportion of the water that contributes to the extra-cellular solution undergoes a transition into ice, the remaining solution becomes increasingly concentrated and thus hypertonic to the cell. To restore the osmotic equilibrium, cellular water will leave the protoplast, resulting in cell dehydration. Generally, freezing rates of 0.5 to 2 °C/min, depending on the type and physiological state of the plant material, are applied. These slow-cooling rates can be obtained using computer-driven cooling devices, stirred methanol baths, and propanol containers held at -70 °C.

6.3.3 Non-penetrating cryoprotective substances

Osmotic dehydration can be obtained through the application of non-penetrating cryoprotective substances, such as sugars, sugar alcohols and high molecular weight additives such as polyethylene glycol (PEG).

6.3.4 Penetrating cryoprotective substances

Commonly used penetrating cryoprotective agents are dimethyl sulphoxide (DMSO) and glycerol. DMSO is preferred for many applications due to its extreme rapid penetration into the cells. Where DMSO toxicity is a problem, glycerol or amino acids (e.g. proline) are often applied.

6.3.5 Adaptive metabolism (hardening)

Hardening is a process that increases the plant's ability to survive the impact of unfavourable environmental stress. In nature this is triggered by environmental parameters, such as reduction in temperature and shortening of day length.

Osmotic changes and abscisic acid (ABA) treatments can also have similar effects. Hardening can result in a considerable increase of proteins, sugars, glycerol, proline and glycine betaine, which all contribute to increasing the osmotic value of the cell solutes.

Most hydrated tissues, however, do not withstand dehydration to moisture contents needed for vitrification (20 to 30 percent) due to solution and mechanical effects. Exceptions are pollen, seeds and somatic embryos of most orthodox seed species. The key for successful cryopreservation thus shifts from freezing tolerance to dehydration tolerance. This tolerance can be induced by chemical cryoprotection with substances like sugars, amino acids, DMSO, glycerol. The mode of action of most of these substances is, however, still far from being understood. Alternatively, tolerance to dehydration can also be induced by adaptive metabolism. For example, it has been observed that cold acclimation in nature often leads to the accumulation of specific proteins, sugars, polyamines and other compounds that can protect cell components during drying. Sugar treatments also result in alterations in protein (Carpentier *et al.*, 2005) and membrane composition (Ramon *et al.*, 2002), the latter influencing both its flexibility and permeability.

6.4 AVAILABLE PLANT CRYOPRESERVATION PROTOCOLS

All cryopreservation protocols described in literature use the above-mentioned techniques or combinations. The most commonly applied protocols are:

6.4.1 Air drying (flash drying, normal drying)

This method is directly applicable to orthodox seed, zygotic embryos and pollen of many common agricultural and horticultural species. Some of these orthodox seeds can even withstand drying below 3 percent moisture content, without any damage and reduction of viability. Flash (or ultra-rapid) drying proved to be beneficial for recalcitrant zygotic embryos of some plant species (Berjak *et al.*, 2000).

6.4.2 Classical slow-cooling (or slow-freezing) protocol

This was the first "standard" protocol developed for hydrated plant tissues (Withers and King, 1980). It is based on slow cooling of specimens (at a rate of 0.5 to 2 °C/min) in the presence of a cryoprotectant solution, generally containing DMSO at a 5 to 15 percent concentration. When during the slow-cooling process a temperature of about -40 °C is reached, the intra-cellular solution is considered to be concentrated enough to vitrify on a subsequent liquid nitrogen plunge. This method is mainly used today for cryopreservation of non-organized tissues, such as cell suspensions and calli.

6.4.3 Encapsulation/dehydration

In this method, developed by Fabre and Dereuddre (1990), explants (usually meristems or embryos) are firstly encapsulated in alginate beads (which can also contain mineral salts and organic substances), thus forming "synthetic seeds" (artificial seeds or synseeds). These synseeds are then treated with a high sucrose concentration, dried down to a moisture content of 20 to 30 percent (under airflow or using silica gel) and subsequently rapidly frozen in liquid nitrogen. Although the procedure can be considered rather lengthy and labour-intensive, it is observed that the presence of a nutritive matrix (the bead) surrounding the explant can promote its regrowth after thawing.

6.4.4 Vitrification

First reports on the use of a vitrification solution with plant tissues appeared in 1989 (Langis *et al.*, 1989; Uragami *et al.*, 1989). The technique relies on treatment of the explants with a concentrated vitrification solution for variable periods of time, from 15 minutes up to two hours, followed by a direct plunge into liquid nitrogen ("vitrification/one-step freezing"). This results in both intra- and extracellular vitrification. The vitrification solution consists of a concentrated mixture of penetrating and non-penetrating cryoprotective substances. The most commonly applied solution, called "Plant Vitrification Solution no. 2" (PVS2), consists of 30 percent glycerol, 15 percent ethylene glycol, 15 percent DMSO (all percentage v/v) and 0.4 M sucrose (Sakai, Kobayashi and Oiyama, 1990). Fifteen years after its first report, vitrification is today by far the most widely used cryopreservation protocol. The success of the procedure can be attributed to its ease, high reproducibility and to the fact that it can successfully be applied to a wide range of tissues and plant species.

6.4.5 Other protocols

Other available methods are the "droplet freezing" (Schäfer-Menuhr, Schumacher and Mix-Wagner, 1997), the "preculture method" (Panis *et al.*, 1996) and the "preculture/dehydration" (Dumet *et al.*, 1993). Up to now these techniques have been applied to only a limited number of plant species and are not described in detail in this review. A recent and promising technique, "encapsulation/vitrification" (Sakai, 2000), is mentioned below in greater detail.

6.5 APPLICATION OF CRYOPRESERVATION TO HERBACEOUS SPECIES

6.5.1 Cell suspensions and callus cultures

Cell suspension and callus cultures are often cryopreserved using the classical slow-cooling protocol. The main aim of cryopreserving these non-organized tissues is not the long-term storage of the genetic diversity, but the conservation of specific features of these tissues that can be lost during normal *in vitro* maintenance. It has been repeatedly reported that the morphogenetic potential of embryogenic callus lines is not affected by their storage in liquid nitrogen. Moreover, it was proven that cryopreservation did not affect the expression of a foreign sam gene in transgenic Papaver somniferum cells (Elleuch *et al.*, 1998), and was beneficial for pyrethrin biosynthesis by *Chrysanthemum cinerariaefolium* cell cultures (Hitmi, Sallanon and Barthomeuf, 1997). Also, the production of regenerable protoplasts is not influenced by cryopreservation, as shown for rice (Jain, Jain and Wu, 1996; Meijer *et al.*, 1991), Festuca and Lolium species (Wang *et al.*, 1994). Moreover, cryopreserved rice (Cornejo, Wong and Blechl, 1995) and maize calli (Gordon-Kamm *et al.*, 1990) proved to be a constant source of regenerable cell cultures for the production of transgenic plants. Moukadiri, Lopes and Cornejo (1999) showed that rice calli, which were stored in liquid nitrogen, had a higher competence for transformation, as indicated by transient gene expression levels. In banana and grape, comparable levels of transient expression, as well as stable transformation, were obtained from cryopreserved and non-cryopreserved suspension cells (Panis *et al.*, 2004; Wang *et al.* 1994). Large-scale cryopreservation of non-organized cultures is reported for coffee (Florin, Brulard and Lepage, 1999), oil palm (Dumet *et al.*, 1994) and banana (Panis *et al.*, 2004).

6.5.2 Cryopreservation of pollen

Pollen is stored for facilitating crosses in breeding programmes, distributing and exchanging germplasm among locations, and preserving nuclear genes of germplasm, as well as for studies in basic physiology, biochemistry, fertility and biotechnology involving gene expression, transformation and *in vitro* fertilization (Towill and Waters, 2000). For example, preservation of pollen can be very useful for cross-pollination of cultivars differing in flowering period. There are methods for the cryopreservation of pollen from many crops (Towill, 1985.), but its application is still rather low and limited to a few research centres (Engelmann, 2004).

6.5.3 Cryopreservation of meristematic tissues

Shoot-meristematic tissues are the most commonly used explants for the cryopreservation of vegetatively propagated species, such as fruit trees and many root and tuber crops. Also, in view of the lower chance for somaclonal variation, organized tissues such as meristems are often preferred over non-organized tissues, such as calli and cell suspensions. Most recent reports deal with encapsulation/dehydration and vitrification methods. It is presumed that in view of its rather severe freeze-dehydration, the classical slow-cooling method is effective in retaining the integrity of individual cells, but less efficient in retaining the tissue integrity necessary for meristem survival. There are many scientific publications on the cryopreservation of meristems ("shoot tips"), but its large-scale application is mainly found in fruit crop germplasm collections (see 6.6.2. "cryopreservation of hardwood trees") in case of herbaceous species, the number of accessions stored in liquid nitrogen is significantly smaller, but continuously growing.

At the German Collection of Micro-Organisms and Cell Cultures (DSMZ, Braunschweig, Germany), meristems of 519 old potato varieties are cryopreserved using the droplet freezing method (Mix-Wagner, Schumacher and Cross, 2003), while at the International Potato Centre (CIP, Lima, Peru), 345 potato accessions are preserved using the vitrification protocol (Panta, personal communication). At K.U. Leuven, Belgium, 306 banana accessions are currently safely stored using the droplet vitrification method (Panis, Piette and Swennen, 2005), representing more than one-quarter of the world's banana collection (INIBAP, 2005). Significant efforts are also made for cassava (Roca *et al.*, 2000), garlic (Kim *et al.*, 2004; Volk, Maness and Rotindo, 2004), mint (Hirai and Sakai, 1999; Towill and Bonnart, 2003) and Australian endangered species (Turner *et al.*, 2001).

6.5.4 Cryopreservation of seeds

Seeds of most common agricultural and horticultural species are tolerant to desiccation and exposure to liquid nitrogen (Stanwood, 1985). Cryogenic storage of orthodox seeds can be considered an alternative to the traditional storage at -20 °C. For some plants species, seed longevity at -20 °C is only a few years and can thus be increased considerably through storage in the vapour phase of liquid nitrogen. In most cases, simple drying to lower the moisture content to 5 to 10 percent is sufficient to resist ultra-low temperatures. Celery is an example of orthodox seed cryopreservation (Gonzalés-Benito *et al.*, 1995).

6.6 APPLICATION OF CRYOPRESERVATION TO WOODY SPECIES

Up to now seed and field collections have been the only reliable option for the long-term germplasm preservation of woody species. However, numerous forest Angiosperms (e.g. *Acer* spp., *Quercus* spp., chestnut, horsechestnut and many tropical species) have non-orthodox seeds with a very limited storage period. Fruit trees, which are mainly vegetatively propagated, require the conservation of huge numbers of accessions in clonal orchards (including old and newly selected cultivars, local varieties and wild material) where a periodic and careful monitoring of the preserved trees is essential. The preservation of the *Prunus* European germplasm, for instance, requires the maintenance of over 30 000 accessions in field repositories spread over 21 countries (Gass, Tobutt and Zanetto, 1996). Hence, cryopreservation should be regarded as a strategically important support to traditional in-field banks. Indeed, the combination of traditional conservation approaches with the potential of cryogenic technology represents an important step forward in minimizing the risks of accidental loss of woody plant genetic resources.

6.6.1 Organs and tissues for woody plant cryopreservation

Recent advances in plant tissue culture have greatly increased the variety of organs and tissues that can be used for storage in liquid nitrogen. Among the explants described above for herbaceous species, three categories are mainly used for woody plant cryopreservation:

(i) *the shoot tips*, differentiated organs which are used for the preservation of vegetatively propagated plants, such as many fruit and timber tree cultivars for which the maintenance of genetic fidelity is fundamental. Shoot tips (1-2 mm, on average) are obtained from apical or axillary buds, excised from *in vitro*-grown shoot cultures. Under sterile conditions, the bud is handled in order to obtain the apical meristem, surrounded by some of the original leaf primordia and leaflets (Niino *et al.*, 1992);

(ii) *the seeds or the isolated embryo axes* for the preservation of species that are mainly reproduced by seeds. For those species characterized by non- or sub-orthodox seeds, which cannot be stored in traditional seed-banks, cryopreservation should be regarded as an important option for the long-term conservation of genetic resources (Pence, 1995);

(iii) *lines of embryogenic callus*, which permit the germplasm conservation of plants with non-orthodox seeds, as well as the maintenance of valuable embryogenic lines utilized in bioengineering, e.g. avoiding the loss of embryogenic potential due to repeated subculturing or allowing the storage of transgenic material while field trials are ongoing.

6.6.2 Cryopreservation of hardwood trees

In recent years, the "vitrification/one-step freezing" technique has been continuously improved and applied to an increasing number of hardwood species. The application of this cryogenic technique to woody species became very popular after the PVS2 solution was introduced for the cryopreservation of *Citrus sinensis* nucellar cells (Sakai, Kobayashi and Oiyama, 1990). To date, the PVS2 solution has been successfully used for the cryopreservation of shoot tips from several economically-important hardwood genera, such as *Malus* (Niino *et al.*, 1992), *Pyrus* (ibid.; Tahtamouni and Shibli, 1999), *Prunus* (Chunnuntapipat *et al.*, 2000; De Carlo, Benelli and Lambardi, 2000; Niino *et al.*, 1997), *Castanea* (Vidal *et al.*, 2005), *Populus* (Lambardi, 2002) and *Vitis* (Matsumoto and Sakai, 2000), easily achieving survival rates higher than 50 percent (Lambardi and De Carlo, 2003). Highest percentages of shoot-tip survival (over 90 percent) are reported for *Prunus jamasakura* (Niino *et al.*, 1997) and *Populus alba* (Lambardi, Fabbri and Caccavale, 2000). It is important to note that when loading explants with the PVS2 solution, post-thaw survival is to a great extent influenced by the duration of treatment, which must be long enough to ensure sufficient cell dehydration without cytotoxic effects. When the solution is applied at 25 °C, exposure time of woody-plant shoot tips ranges from 20 (Matsumoto *et al.*, 2001) to 120 min (Niino *et al.*, 1997; Vidal *et al.*, 2005). Alternatively, to reduce the risk of toxicity, a chilled solution (0 °C) can be used (Caccavale, Lambardi and Fabbri, 1998; Valladares *et al.*, 2004), or a gradient of concentration is applied (e.g. 50 percent for 30 min followed by 100 percent for 50 min) (Matsumoto and Sakai, 2000). Following the storage in liquid nitrogen, rapid warming in a waterbath is required. This will avoid recrystallization and ensures a proper recovery of the vitrified material. Thawing temperatures ranging from 20 °C to 45 °C have been proposed for woody species, 40 °C (corresponding approximately to a warming rate of 180 °C/min) being by far the most commonly applied (Lambardi and De Carlo, 2003).

In addition, the "encapsulation/dehydration" technique has repeatedly shown itself to be very effective for the long-term conservation of hardwood species. Effective protocols have already been developed for several important species, such as *Malus*, *Pyrus* and *Prunus* spp., with post-thaw shoot-tip survival and regrowth rates often exceeding 80 percent (Lambardi and De Carlo, 2003).

"Encapsulation/vitrification" is a new technique (Sakai, 2000), combining the encapsulation of explants with the application of a vitrification mixture. It has already been successfully applied to apple (Paul, Daigny and Sangwan-Norreel, 2000) and plum (De Carlo, Benelli and Lambardi, 2000) shoot tips. On the other hand, it showed to be less effective than a classic "vitrification/one-step freezing" approach for the long-term preservation of juvenile shoot tips of common ash (Schoenweiss, Meier-Dinkel and Rüdiger, 2005) showing that further improvement of protocols is necessary.

Recently, the cryopreservation method for dormant-vegetative buds (winter buds), based on the original procedure described by Sakai (1960), was applied to 1 915 apple accessions (Towill *et al.*, 2004). Accordingly, winter-collected scions were desiccated in a cold room at -5 °C to 30 percent moisture and cooled down slowly (1 °C per hour) to -30 °C, where they remained for 24 hours prior to being transferred to the vapour phase of liquid nitrogen (-160 °C). After retrieval from storage and grafting onto rootstocks, over 90 percent of the accessions showed a survival higher than 30 percent. Similarly, a successful multi-step pre-freezing procedure was developed for mulberry and pear winter buds. Here, the hardening treatment consisted in a 5 °C daily reduction of bud temperature to bring it down from 0 °C to -30 °C, followed by direct immersion in liquid nitrogen (Niino, 2000). A more classic "vitrification/one-step freezing" approach has been proposed for the cryopreservation of persimmon (*Diospyros kaki*) winter-dormant axillary buds, directly collected from the field (Matsumoto *et al.*, 2001).

Compared to shoot tips, reports dealing with the cryopreservation of embryogenic calli and somatic embryos are limited for hardwood trees. The number of temperate hardwood species for which a "one-step freezing" procedure has been developed has increased in recent years include *Castanea sativa* (Correidoira *et al.*, 2004), *Fraxinus angustifolia* (Tonon *et al.*, 2001), *Quercus suber* (Valladares *et al.*, 2004), *Olea europaea* (Lambardi *et al.*, 2002) and *Aesculus hippocastanum* (Lambardi, De Carlo and Capuana, 2005.). In addition, embryogenic calli of some tropical forest and fruit species have been successfully cryopreserved (Sudarmonowati, 2000; Wu *et al.*, 2003). In the above-mentioned reports, isolated somatic embryos or pieces of embryogenic callus are used as explants and treated with PVS2 or, alternatively, encapsulated and dehydrated prior to a direct plunge into liquid nitrogen.

For hardwood species, in addition to the direct immersion of specimens in liquid nitrogen, the use of the slow-cooling technique is still sporadically reported. With the slow cooling approach, shoot-tip survival is variable, ranging from a minimum of 34 percent (*Juglans regia* [de Boucaud and Brison, 1995]) up to a maximum of 92 percent (*Malus* spp. [Zhao *et al.*, 1999]). Also, a combination of "encapsulation/dehydration" and slow cooling has been proposed for the cryopreservation of walnut somatic embryos (de Boucaud, Brison and Negrier, 1994).

6.6.3 Cryopreservation of softwood trees

Cryopreservation plays an important role in the clonal selection programmes of conifers, based on somatic embryogenesis. Effective procedures based upon the slow-cooling approach have been successfully developed for embryogenic cultures of numerous species of *Picea, Pinus, Larix, Abies* and *Pseudotsuga*. An important programme for the cryopreservation of conifer germplasm has been developed in British Columbia, where over 5 000 embryogenic lines are presently stored in liquid nitrogen (Cyr, 2000) At present, no reports are available dealing with cryopreservation of conifer shoot tips.

6.6.4 Seed and embryonic axis cryopreservation

As concerns the long-term storage of seeds, effective cryopreservation procedures have been developed for both whole seeds and excised embryonic axes from many temperate (Pence, 1995) and tropical (Marzalina and Krishnapillay, 1999) woody species. Pre-dessiccation to a moisture content below 20 percent, followed by slow cooling or by direct immersion in liquid nitrogen, is the common methodology applied to orthodox seeds. However, very promising results have also been obtained with woody species characterized by sub-orthodox or non-orthodox (recalcitrant) seeds. In polyembryonic Citrus and related species, for instance, various procedures have been developed for the cryostorage of their seeds (Cho *et al.*, 2001; Cho et al, 2002; Rhadamani and Chandel, 1992), and the regrowth of both zygotic and nucellar embryos has been reported (Lambardi *et al.*, 2004). Both kinds of embryos are very important for evolution, breeding, and propagation of Citrus species.

6.6.5 Cryostored woody plant germplasm

Today, promising examples of cryogenic repositories for woody species are available. In addition to the aforesaid collection of conifer embryogenic lines in British Columbia, the following cryo-banks can be mentioned (Reed, 2001):

- the National Seed Storage Laboratory (NSSL) of Fort Collins (USA), with around 2 100 accessions of apple (dormant buds);
- the National Clonal Germplasm Repository (NCGR) of Corvallis (USA), with over 100 accessions of pear (shoot tips);
- Association Forêt-Cellulose (AFOCEL) of France, with over 100 accessions of elm (dormant buds);
- the National Institute of Agrobiological Resources (NIAR) of Japan, with around 50 accessions of mulberry.

Some tropical and sub-tropical woody species are also now cryopreserved, e.g. at Institut de Recherche pour le développement (IRD) of France (80 accessions of oil palm [Engelmann, 2004]), and at the National Bureau of Plant Genetic Resources (NBPGR), India (numerous accessions of citrus, jackfruit, almond, litchi and tea [Reed, 2001]).

6.7 GENETIC INTEGRITY OF PLANTS FROM CRYOPRESERVATION

The occurrence of somaclonal variation has been reported several times as a consequence of the introduction, manipulation and regeneration of plants under *in vitro* conditions. Since cryopreservation involves many *in vitro* steps, careful attention must be given to avoid the incidence of somaclonal variation before and after the conservation of germplasm in liquid nitrogen. On the other hand, during storage the main peculiarities of the cryopreservation technology (e.g. the blocked metabolism of cells and the absence of subcultures) reduce the risks of genetic and epigenetic alterations to nil. Some threats to genetic stability arise from particular reactions (free radical formation, molecular damage due to ionizing radiation) that might still occur at the temperature of -196 °C (Grout, 1990) as well as from the common practice of using DMSO as cryoprotectant at concentrations up to 10 percent. Although the number of reports studying these aspects in detail are still limited, the fact that up to now no clear evidence of morphological, cytological or genetic alterations due to cryopreservation has been produced is promising. (For a thorough review, see Harding, 2004.)

6.8 CONCLUSIONS

Although the slow-cooling approach was already introduced in the 1970s, for a long time cryopreservation of plant tissues was not studied on a wide scale. This was mainly due to the complexity of procedures and the high cost of computer-driven cryo-freezers. With the development in the early 1990s of "new" and more simplified cryopreservation protocols, based on the prevention of intra- and extra-cellular ice crystals by means of cell dehydration followed by direct immersion of explants in liquid nitrogen, the cryostorage of genetic resources has become a realistic target for many plant species. Nowadays, although the vitrification protocol could be considered a "standardized" protocol, a large amount of the work is still performed in the framework of academic studies and involves only one or a few accessions per plant species. This is mainly due to the complex and time-consuming optimization of procedures that is always required before an efficient cryopreservation protocol is established for a new species or accession. The two most important parameters that need to be optimized for each species and tissue are the preparation phase of tissues for the dehydration phase (most important are sugar and/or cold treatments) and the length of exposure to the vitrification solution (or, in the case of encapsulated explants, to air flow or silica gel). Research should move in the direction of simplifying and standardizing the procedures as much as possible in order to make the technology available to a wide range of public institutions and private companies. Moreover, to facilitate the development of even more efficient cryopreservation protocols, a better knowledge of the physico-chemical background of cryopreservation is needed. This can only be obtained through fundamental studies that involve both thermal analysis and a thorough examination of the different parameters that can influence the cryo-behaviour, such as endogenous sugars, membrane composition, oxidative stress and cryoprotective proteins. These parameters are now investigated for different plant species in the framework of the European FP5 project, Establishing Cryopreservation Methods for Conserving European Plant Germplasm Collections (CRYMCEPT). (See www.agr.kuleuven.ac.be/dtp/tro/CRYMCEPT.)

6.9 ACKNOWLEDGEMENTS

B. Panis gratefully acknowledges the financial support of the Directorate General of International Collaboration (DGIC), Belgium, the International Network for the Improvement of Banana and Plantain (INIBAP) within the framework of the Genetic Improvement Group of the Global Programme for Musa Improvement (PROMUSA). Part of this study has also been carried out with financial support from the Commission of the European Communities' specific cooperative research programme, Quality of Life and Management of Living Resources, QLK5-2002-1279, CRYMCEPT. This study does not necessarily reflect the Commission's views and in no way anticipates its future policy in this area. M. Lambardi's research activity is financially supported by the National Research Council (CNR) of Italy, Agri-food Department Project, Risorse biologiche e tutela dell'agroecosistema. Special thanks are due to Carla Benelli and Anna De Carlo for their valuable and competent participation in the Cryopreservation Working Group of the Istituto per la valorizzazione del legno e delle specie arboree (IVALSA).

6.10 REFERENCES

Benson, E.E. 1999. Cryopreservation. *In* E.E. Benson, ed. *Plant Conservation Biotechnology.* London, England, Taylor & Francis. pp. 83-95.

Berjak, P., Walker, M., Mycock, D.J., Wesley-Smith, J., Watt, P. & Pammenter, N.W. 2000. Cryopreservation of recalcitrant zygotic embryos. *In* F. Engelmann & H. Takagi, eds. *Cryopreservation of tropical plant germplasm.* Rome, International Plant Genetic Resources Institute. pp. 140-155.

Botanic Gardens Conservation International (BGCI). 2005. (available at www.bgci.org)

Brison, M., de Boucaud, M.T., Pierronnet, A. & Dosba, F. 1997. Effect of cryopreservation on the sanitary state of a *Prunus* rootstock experimentally contaminated with Plum Pox Potyvirus. *Plant Sci.*, 123: 189-196.

Caccavale, A., Lambardi, M. & Fabbri, A. 1998. Cryopreservation of woody plants by axillary bud vitrification: a first approach with poplar. *Acta Hortic*, 457: 79-83.

Carpentier, S., Witters, E., Laukens, K., Deckers, P., Swennen, R. & Panis, B. 2005. Preparation of protein extract from recalcitrant plant tissues: an evaluation of different methods for two-dimensional gel electrophoresis analysis. *Proteomics*, 5: 2497-2507.

Cho, E.G., Hor, Y.L., Kim, H.H., Rao, V.R. & Engelmann, F. 2001. Cryopreservation of *Citrus madurensis* zygotic embryonic axes by vitrification: importance of pregrowth and preculture conditions. *CryoLetters*, 22: 391-396.

Cho, E.G., Normah, M.N., Kim, H.H., Rao, V.R. & Engelmann, F. 2002. Cryopreservation of *Citrus aurantifolia* seeds and embryonic axes using a dessiccation protocol. *CryoLetters*, 23: 309-316.

Chunnuntapipat C., Collins G., Bertozzi T. & Sedgley, M. 2000. Cryopreservation of *in vitro* almond shoot tips by vitrification. *J. Hort. Sci. & Biotech.*, 75: 228-232.

Cornejo, M.J., Wong, V.L. & Blechl, A.E. 1995. Cryopreserved callus: a source of protoplasts for rice transformation. *Plant Cell Rep.*, 14: 210-214.

Correidoira, E., San José, M.C., Ballester, A. & Vieitez, A.M. 2004. Cryopreservation of zygotic embryo axes and somatic embryos of European chestnut. *CryoLetters*, 25: 33-42.

Cyr, D.R. 2000. Cryopreservation: roles in clonal propagation and germplasm conservation of conifers. *In* F. Engelmann and H. Takagi, eds. *Cryopreservation of tropical plant germplasm.* Rome, International Plant Genetic Resources Institute. pp. 261-268.

de Boucaud, M.T. & Brison M. 1995. Cryopreservation of germoplasm of walnut (*Juglans* species). *In* Y.P.S. Bajaj, ed. Cryopreservation of plant germplasm I. *Biotechnology in agriculture and forestry*, Vol. 32. Berlin, Heidelberg, New York, USA, Springer. pp. 129-147.

de Boucaud, M.T., Brison, M. & Negrier, P. 1994. Cryopreservation of walnut somatic embryos. *CryoLetters*, 15: 151-160.

De Carlo, A., Benelli, C. & Lambardi, M. 2000. Development of a shoot-tip vitrification protocol and comparison with encapsulation-based procedures for plum (*Prunus domestica* L.) cryopreservation. *CryoLetters*, 21: 215-222.

Dumet, D., Engelmann, F., Chabrillange, N. & Duvall, Y. 1993. Cryopreservation of oil palm (*Elaeis guinensis* Jacq.) somatic embryos involving a desiccation step. *Plant Cell Rep.*, 12: 352-355.

Dumet, D., Engelmann, F., Chabrillange, N. & Duvall, Y. 1994. Effect of desiccation and storage temperature on the conservation of cultures of oil palm somatic embryos. *CryoLetters*, 15: 85-90.

Elleuch, H., Gazeau, C., David, H. & David, A. 1998. Cryopreservation does not affect the expression of a foreign sam gene in transgenic *Papaver somniferum* cells. *Plant Cell Rep.*, 18: 94-98.

Engelmann, F. 2004. Plant cryopreservation: progress and prospects. *In vitro Cell. & Dev. Biol.-Plant*, 40: 427-433.

Fabre, J. & Dereuddre, J. 1990. Encapsulation-dehydration: a new approach to cryopreservation of *Solanum* shoot tips. *CryoLetters*, 11: 413-126.

Florin, B., Brulard, E. & Lepage, B. 1999. Establishment of a cryopreserved coffee germplasm bank. *Cryo'99, World Congress of Cryobiology*. Marseille, France, 12-15 July 1999, p. 167.

Gass, T., Tobutt, K.R. & Zanetto, A. 1996. *Report of the Working Group on Prunus.* Fifth meeting. Rome, International Plant Genetic Resources Institute. pp. 1-70.

Gonzalés-Benito, M.E., Iriondo, J.M., Pita, J.M. & Perez-Garcia, F. 1995. Effects of seed cryopreservation and priming on germination in several cultivars of *Apium graveolens. Ann. Bot.*, 75: 1-4.

Gordon-Kamm, W.J., Spencer, T.M., Mangano, M.L., Adams, T.R., Daines, R.J., Start, W.G., O'Brien, J.V., Chambers, S.A., Adams, W.R., Willetts, N.G., Rice, T.B., MacKey, C.J., Krueger, R.W., Kausch, A.P. & Lemaux, P.G. 1990. Transformation of maize cells and regeneration of fertile transgenic plants. *Plant Cell*, 2: 603-618.

Grout, B. 1990. *In vitro* conservation of germplasm. *In* S.S. Bhojwani, ed. *Developments in crop science, Vol. 19. Plant tissue culture: applications and limitations.* Elsevier, The Netherlands. pp. 394-410.

Harding, K. 2004. Genetic integrity of cryopreserved plant cells: a review. *CryoLetters*, 25: 3-22.

Helliot, B., Panis, B., PouMay, Y., Swennen, R., Lepoivre, P. & Frison, E. 2002. Cryopreservation for the elimination of cucumber mosaic and banana streak viruses from banana (*Musa* spp.). *Plant Cell Rep.*, 20: 1117-1122.

Hirai, D. & Sakai, A. 1999. Cryopreservation of *in vitro*-grown axillary shoot-tip meristems of mint (*Mentha spicata* L.) by encapsulation vitrification. *Plant Cell Rep.*, 19: 150-155.

Hitmi, A., Sallanon, H. & Barthomeuf, C. 1997. Cryopreservation of *Chrysanthemum cinerariaefolium* Vis. cells and its impact on their pyrethrin biosynthesis ability. *Plant Cell Rep.*, 17: 60-64.

INIBAP. 2005. *INIBAP annual report 2004.* International Network for the Improvement of Banana and Plantain, Montpellier, France. (also available at www.inibap.org/pdf/2004ra_en.pdf).

IPGRI. 2004. (available at www.ipgri.cgiar.org/themes/human/economics.htm).

Jain, S., Jain, R.K. & Wu, R. 1996. A simple and efficient procedure for conservation of embryogenic cells of aromatic Indica rice varieties. *Plant Cell Rep.*, 15: 712-717.

Kim, H.H., Cho, E.G., Baek, H.J., Kim, C.Y., Keller, E.R.J. & Engelmann, F. 2004. Cryopreservation of garlic shoot tips by vitrification: effects of dehydration, rewarming, unloading and regrowth conditions. *CryoLetters*, 25: 59-70.

Lambardi, M. 2002. Cryopreservation of Germplasm of Populus (Poplar) Species. *In* L. Towill & Y.P.S. Bajaj, eds. *Cryopreservation of plant germplasm II. Biotechnology in agriculture and forestry*, Vol. 50. Berlin Heidelberg, Germany, Springer. pp. 269-286.

Lambardi, M., Fabbri A. & Caccavale, A. 2000. Cryopreservation of white poplar (*Populus alba* L.) by vitrification of *in vitro*-grown shoot tips. *Plant Cell Rep.*, 19: 213-218.

Lambardi, M., Lynch, P.T., Benelli, C., Mehra, A. & Siddika, A. 2002. Towards the cryopreservation of olive germplasm. *Adv. Hort. Sci.*, 16(3-4): 165-174.

Lambardi, M. & De Carlo, A. 2003. Application of tissue culture to the germplasm conservation of temperate broad-leaf trees. *In* S.M. Jain & K. Ishii, eds. *Micropropagation of woody trees and fruits.* Dordrecht, Holland, Kluwer Ac. Pub. pp. 815-840.

Lambardi, M., De Carlo, A., Biricolti, S., Puglia, A.M., Lombardo, G., Siragusa, M. & De Pasquale, F. 2004. Zygotic and nucellar embryo survival following dehydration/cryopreservation of Citrus intact seeds. *CryoLetters*, 25: 81-90.

Lambardi, M., De Carlo, A. & Capuana, M. 2005. Cryopreservation of embryogenic callus of *Aesculus hippocastanum* L. by vitrification/one-step freezing. *CryoLetters*, 26: 185-192.

Langis, R., Schnabel, B., Earle, E.D. & Steponkus, P.L. 1989. Cryopreservation of *Brassica campestris* L. cell suspensions by vitrification. *CryoLetters*, 10:421-428.

Marzalina, M. & Krishnapillay, B. 1999. Recalcitrant seed biotechnology application to rain forest conservation. *In* Erica E. Benson, ed. *Plant conservation biotechnology.* Taylor & Francis, London, England. pp. 265-276.

Matsumoto, T. & Sakai, A. 2000. Cryopreservation of grape *in vitro*-cultured axillary shoot tips by three-step vitrification. *In* F. Engelmann & H. Takagi, eds. *Cryopreservation of tropical plant germplasm.* Rome, IPGRI. pp. 424-425.

Matsumoto T., Mochida K., Itamura H. & Sakai A. 2001. Cryopreservation of persimmon *(Diospyros kaki Thunb.)* by vitrification of dormant shoot tips. *Plant Cell Rep.*, 20: 398-402.

Meijer, E.G.M., Vaniren, F., Schrijnemakers, E., Hensgens, L.A.M., Vanzijderveld, M. & Schilperoort, R.A. 1991. Retention of the capacity to produce plants from protoplasts in cryopreserved cell lines of rice *(Oryza sativa L.). Plant Cell Rep.*, 10: 171-174.

Mix-Wagner, G., Schumacher, H.M. & Cross R.J. 2003. Recovery of potato apices after several years of storage in liquid nitrogen. *CryoLetters*, 24: 33-41.

Moukadiri O., Lopes, C.R. & Cornejo, M.J. 1999. Physiological and genomic variations in rice cells recovered from direct immersion and storage in liquid nitrogen. *Physiol. Plant.* 105: 441-449.

Niino, T. 2000. Cryopreservation of deciduous fruits and mulberry trees. *In* M.K. Razdan & E.C. Cocking, ed. *Conservation of plant genetic resources in vitro.* Enfield, USA, Science Pub. Inc. pp 195-223.

Niino, T., Sakai, A., Yakuwa, H. & Nojiri, K. 1992. Cryopresevation of *in vitro*-grown shoot tips of apple and pear by vitrification. *Plant Cell Tiss. Org. Cult.*, 28: 261-266.

Niino, T., Tashiro, K., Suzuki, M., Ohuchi, S., Magoshi, J. & Akihama, T. 1997. Cryopreservation of *in vitro* shoot tips of cherry and sweet cherry by one-step vitrification. *Sci. Hort.*, 70: 155-163.

Panis, B., Totté, N., Van Nimmen, K., Withers, L.A. & Swennen, R. 1996. Cryopreservation of banana (*Musa* spp.) meristem cultures after preculture on sucrose. *Plant Sci.*, 121: 95-106.

Panis, B., Strosse, H., Remy, S., Sági, L. & Swennen, R. 2004. Cryopreservation of banana tissues: support for germplasm conservation and banana improvement. *In* S.M. Jain & R. Swennen, eds. *Banana improvement: cellular, molecular biology, and induced mutations.* Enfield, USA, Science Publishers Inc. pp. 13-21.

Panis, B., Piette, B. & Swennen R. 2005. Droplet vitrification of apical meristems: a cryopreservation protocol applicable to all Musaceae. *Plant Sci.*, 168: 45-55.

Paul, H., Daigny, G. & Sangwan-Norreel, B.S. 2000. Cryopreservation of apple (*Malus x domestica* Borkh.) shoot tips following encapsulation-dehydration or encapsulation-vitrification. *Plant Cell Rep.*, 19: 768-774.

Pence, V.C. 1995. Cryopreservation of recalcitrant seeds. *In* Y.P.S. Bajaj, ed. *Cryopreservation of plant germplsm I. Biotechnology in agriculture and forestry,* vol. 32. Berlin Heidelberg, Germany, Springer. pp. 29-50.

Ramon, M., Geuns, J., Swennen, R. & Panis, B. 2002. Polyamines and fatty acids in sucrose precultured banana meristems and correlation with survival rate after cryopreservation. *CryoLetters*, 23: 345-352.

Reed, B.M. 2001. Implementing cryogenic storage of clonally propagated plants. *CryoLetters*, 22: 97-104.

Rhadamani, J. & Chandel, K.P.S. 1992. Cryopreservation of embryonic axes of trifoliate orange (*Poncirus trifoliata* [L.] RAF.). *Plant Cell Rep.*, 11: 372-374.

Roca, W., Debouck, D., Escobar, R. & Mafla, G. 2000. Cryopreservation and cassava germplasm conservation at CIAT. *In* F. Engelmann & H. Takagi, eds. *Cryopreservation of tropical plant germplasm.* International Plant Genetic Resources Institute, Rome. pp. 273-279.

Sakai, A. 1960. Survival of the twigs of woody plants at -196 °C. *Nature*, 185: 393-394.

Sakai, A. 2000. Development of cryopreservation techniques. *In* F. Engelmann & H. Takagi, eds., *Cryopreservation of tropical plant germplasm.* International Plant Genetic Resources Institute, Rome, pp. 1-7.

Sakai A., Kobayashi, S. & Oiyama, I. 1990. Cryopreservation of nucellar cells of navel orange (*Citrus sinensis* Osb. var. *brasiliensis* Tanaka) by vitrification. *Plant Cell Rep.*, 9: 30-33.

Schäfer-Menuhr, A., Schumacher, H.M. & Mix-Wagner G. 1997. Cryopreservation of potato cultivars - design of a method for routine application in genebanks. *Acta Hortic.*, 447: 477-482.

Schoenweiss, K., Meier-Dinkel, A., Rüdiger, G. 2005. Comparison of cryopreservation techniques for long-term storage of ash *(Fraxinus excelsior L.).* *CryoLetters,* 26: 201-212.

Stanwood, P.C. 1985. Cryopreservation of seed germplasm for genetic conservation. *In* K.K. Kartha, ed. *Cryopreservation of plant cells and organs.* Boca Raton, Florida, CRC Press. pp. 199-226.

Sudarmonowati, E. 2000. Cryopreservation of tropical plants: current research status in Indonesia. *In* F. Engelmann & H. Takagi, eds. *Cryopreservation of tropical plant germplasm.* Rome, International Plant Genetic Resources Institute. pp. 291-296.

Tahtamouni, R.W. & Shibli, R.A. 1999. Preservation at low temperature and cryopreservation in wild pear *(Pyrus syriaca). Adv. Hort. Sci.,* 13: 156-160.

Tonon, G., Lambardi, M., De Carlo, A. & Rossi, C. 2001. Crioconservazione di linee embriogeniche di *Fraxinus angustifolia* Whal. *Proceedings of the VI Convegno Nazionale Biodiversità* (with English abstract). Bari, Italy, 6-7 September 2001. pp. 619-625.

Towill L.E. 1985. Low temperature and freeze-vacuum-drying preservation of pollen. *In* K.K. Kartha, ed. *Cryopreservation of plant cells and organs.* Boca Raton, Florida, CRC Press. pp. 171-198.

Towill, L.E. & Waters, C. 2000. Cryopreservation of pollen. *In* F. Engelmann & H. Takagi, eds. *Cryopreservation of Tropical Plant Germplasm.* Rome, International Plant Genetic Resources Institute. pp. 115-129.

Towill L.E. & Bonnart R. 2003. Cracking in a vitrification solution during cooling or warming does not affect growth of cryopreserved mint shoot tips. *CryoLetters,* 24: 341-346.

Towill, L.E., Forsline, P.L., Walters, C., Waddell, J.W. & Laufmann, J. 2004. Cryopreservation of *Malus* germplasm using a winter vegetative bud method: results from 1915 accessions. *CryoLetters,* 25: 323-334.

Turner, S.R., Senaratna, T., Bunn, E., Tan, B., Dixon, K.W. & Touchell, D.H. 2001. Cryopreservation of shoot tips from six endangered Australian species using a modified vitrification protocol. *Ann. Bot.,* 87: 371-378.

Uragami, A., Sakai, A., Nagai, M. & Takahashi, T. 1989. Survival of cultured cells and somatic embryos of *Asparagus officinalis* L. cryopreserved by vitrification. *Plant Cell Rep.,* 8: 418-421.

Uragami A., Sakai A. & Nagai M. 1990. Cryopreservation of dried axillary buds from plantlets of *Asparagus officinalis* L. grown *in vitro. Plant Cell Rep.,* 9: 328-331.

Valladares, S., Toribio, M., Celestino, C. & Vietez, A.M. 2004. Cryopreservation of embryogenic cultures from mature *Quercus suber* trees using vitrification. *CryoLetters,* 25: 177-186.

Vidal, N., Sanchez, C., Jorquera, L., Ballester, A. & Vieitez, A.M. 2005. Cryopreservation of chestnut by vitrification of *in vitro*-grown shoot tips. *In vitro Cell. & Dev. Biol.-Plant,* 41: 63-68.

Volk, G.M., Maness, N. & Rotindo, K. 2004. Cryopreservation of garlic *(Allium sativum L.)* using Plant Vitrification Solution 2. *CryoLetters,* 25: 219-226.

Wang Z.Y., Legris G., Nagel J., Potrykus I. & Spangenberg G. 1994. Cryopreservation of embryogenic cell suspensions in *Festuca* and *Lolium* species. *Plant Sci.,* 103: 93-106.

Wang, Q.C., Marassi, M., Li, P., Gafny, R., Sela, I. & Tanne, E. 2003. Elimination of grapevine virus A (GVA) by cryopreservation of *in vitro*-grown shoot tips of *Vitis vinifera* L. *Plant Sci.,* 165: 321-327.

Withers, L.A. & King, P.J. 1980. A simple freezing unit and routine cryopreservation method for plant cell cultures. *CryoLetters,* 1: 213-220.

Wu, Y., Huang, X., Xiao, J., Li, X., Zhou, M. & Engelmann, F. 2003. Cryopreservation of mango *(Mangifera indica L.)* embryogenic cultures. *CryoLetters*, 24: 303-314.

Zhao, Y., Wu Y., Engelmann, F., Zhou M. & Chen, S. 1999. Cryopreservation of apple *in vitro* shoot tips the droplet freezing method. *CryoLetters*, 20: 109-112.

III. Use of molecular markers for characterization and conservation of genetic resources

7. Use of molecular markers and other information for sampling germplasm to create an animal gene bank

Henner Simianer

7.1 INTRODUCTION

The need to conserve farm animal biodiversity is accepted by many countries through the ratification of the Convention of Biological Diversity (CBD, 1992). Although farm animal breeds are not well defined in general terms, they are the typical unit used for conservation. Based on an inventory of the actual breeds carried out on a global scale by FAO through the World Watch List (FAO, 2000), the specific breed characteristics and their genetic diversity, decisions need to be taken on what should be conserved.

Figure 11 gives an overview of the different conservation schemes available for farm animals. It is generally accepted that whenever possible, preference should be given to *in vivo* conservation schemes; *in vitro* schemes are often started as an additional safeguard when the population size of the endangered breeds is very small or the breed is at high risk of extinction. Semen and embryos are the first choice for cryo-conservation. It should be noted, however, that while long-term storage of deep-frozen semen is feasible in all major farm animal species, the use of cryo-conserved embryos is only established in cattle and small ruminants and under development in a number of other species (Hall, 2004). Now that Dolly has been successfully cloned, storing somatic cells to produce clones in the future may be

FIGURE 11
Systematic overview of basic conservation schemes for farm animals

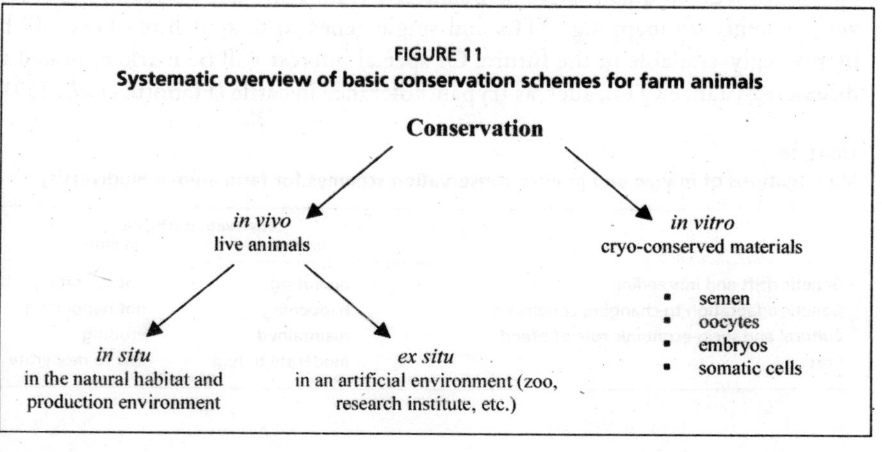

seen as a simple and inexpensive option (Wilmut *et al.*, 1997), but up to now practical experience with this technique is rather limited in most species.

Table 10 shows some of the main features of *in vivo* and *in vitro* conservation schemes. It should be noted that the apparent advantage that a live population genetically adapts to changing conditions is often overemphasized since genetic change due to natural selection is not expected to be very large in only a few generations. The costs of *in vitro* schemes are often considered very low. If costs of reactivating the cryo-conserve at the end of the planning horizon are included, however, this may not be true. Reist-Marti (2004) pointed out that even the collecting and storing phase of a cryo-conservation scheme will be expensive if the technical infrastructure and the expertise are not available, which is the case in most developing countries.

Creating an animal gene bank may refer not only to *in vitro*, but also to *in vivo* repositories; in the latter case it might be understood as identifying a sub-population that is actively managed in a conservation breeding programme, for example, through a sire rotation system (Kimura and Crow, 1963).

7.2 USE OF MOLECULAR MARKERS

Molecular markers are a tool to study diversity on the genetic level. The most widespread use of molecular markers in this context is the assessment of diversity within and between breeds. Although in principle all types of markers would be suitable for this purpose, microsatellites are used in 90 percent of all diversity studies (Baumung, Simianer and Hoffmann, 2004). A joint committee of FAO and the International Society for Animal Genetics (ISAG) has recommended a standard set of microsatellite markers for the major farm animal species. This recommendation was recently reviewed and extended to a larger number of species (Hoffman *et al.*, 2004). These markers are chosen to represent neutral genetic variability in the genome.

In addition, one might also consider markers associated with so-called quantitative trait loci (QTL), i.e. markers that reflect the genetic potential of an animal for a given quantitative or qualitative trait. Farm animal research focuses very strongly on mapping QTLs and single genes so that such markers will be increasingly available in the future. Of special interest will be markers linked to disease-resistance QTL, such as trypanotolerance in cattle (Hanotte *et al.*, 2003),

TABLE 10
Main features of *in vivo* and *in vitro* conservation schemes for farm animal biodiversity

	Conservation scheme	
	in vivo	*in vitro*
Genetic drift and inbreeding	operating	not operating
Genetic adaptation to changing conditions	happening	not happening
Cultural and socio-economic role of breed	maintained	eroding
Cost	moderate to high	low to moderate

nematode resistance in sheep (Coltman *et al.*, 2001), and E.coli-resistance in pigs (Meijerink *et al.*, 2000).

7.3 CREATING A GENE BANK

Creating a gene bank can be considered a multi-stage decision-making process with the following stepwise decisions:

7.3.1 Which breeds to conserve?

Weitzman (1992; 1993) has presented a formal framework for the decision-making process on breeds to conserve. While the diversity metric suggested by Weitzman has been criticized as not accounting for within-breed diversity (see, for example, Caballero and Toro, 2002), the general framework to make decisions based on the expected conserved diversity has a strong appeal. The original proposition, which was first adopted by Thaon d'Arnoldi, Foulley and Ollivier (1998) for farm animals, is based on a diversity metric derived from a genetic distance matrix. Genetic distances can be estimated from allele frequencies at marker loci between populations: the greater the difference in these frequencies, the more distant the breeds are. Diversity is a measure for the total variability of a distance matrix for a set of breeds: the more distant the breeds in the set are, the larger the diversity will be.

Using actual allele frequencies results in the present diversity. Combining the information with extinction probabilities, i.e. the probability that a breed will go extinct over a given time horizon, for example, 50 years, results in the expected diversity at the end of the time horizon. The expected diversity will always be less than the actual diversity. The objective is to design conservation programmes in such a way that the expected diversity is maximized.

Weitzman (1993) suggests that the conservation potential is the single most informative criterion to rank breeds with respect to conservation priority. The conservation potential of a breed basically reflects the amount of expected diversity that can be conserved if a breed is made completely safe. Simianer *et al.* (2003) found that this criterion correctly identified the optimum subset of six breeds to be conserved in a set of 23 African cattle breeds. Pinent, Simianer, and Weigend (2005) show that the derivation of the conservation potential for a set of breeds should always take into account information on related breeds outside the set of candidate breeds for conservation (e.g. foreign breeds or commercial breeding populations) to avoid "false positives".

Simianer (2002; 2005) suggests combining expected diversity with other criteria resulting in the expected total utility as a maximization criterion. This criteria may encompass the presence of special genetic traits (such as disease tolerance), production, and cultural or environment values of breeds, *inter alia*. A similar but less formal argument was made by Piyasatian and Kinghorn (2003).

7.3.2 Optimal allocation of resources

Once the decision is made as to which breeds should be sampled, it is necessary to assign appropriate shares of the conservation budget to the different breeds. In the European Union, about 40 million euros are spent per year for the conservation of farm animal genetic resources (Signorello and Pappalardo, 2003). Although this is not a centralized budget distributed in a uniform and rational process, in principle one could compare the theoretically optimal allocation with the real situation, revealing the relative efficiency of the implemented conservation programmes.

Simianer *et al.* (2003) have suggested a formal approach to find the optimal allocation of a given budget. This basic approach was further refined by Reist-Marti (2004). In a first application, Simianer (2002) showed that the optimal allocation of a hypothetical conservation budget to a set of 26 African cattle breeds resulted in a 60 percent increase of efficiency (in terms of conserved diversity per conservation funds) compared to uniform distribution or allocation of the total budget to the three most endangered breeds only.

7.3.3 Conservation scheme criteria for a given breed

This decision is not independent from the choice of breeds to conserve and the optimal allocation of resources. To do these first steps, costs and effects (in terms of reduced extinction probability) need to be known for the different conservation schemes. The costs can typically be subdivided into fixed costs, which are necessary to establish the conservation scheme in this breed, and variable costs, which depend on the number of animals, herds, cryo-conserved sample, *inter alia*. With known cost functions for different conservation schemes in the same breed, it is always possible to identify the optimum conservation scheme for a given investment level within breed.

This is demonstrated in Reist-Marti (2004) where three out of four different conservation schemes were found to be preferable in at least one out of eight African cattle breeds chosen for conservation. If such a planning process is considered on an international level, factors such as the exchange rate of currencies and relative labour costs in different countries play an important role. Labour-intensive *ex situ* conservation schemes may be cheaper than cryo-conservation in some countries where wages are low and the infrastructure for cryo-conservation is not available (Reist-Marti, 2004).

7.3.4 Which germplasm should be stored?

Once the aforementioned decisions are made, individual genotypes need to be identified to become part of the conservation scheme. Some general criteria can be defined concerning the desirable genetic properties of the sample:

- It should represent the genetic portfolio of the breed.
- It should have a maximum effective population size.
- Special genetic traits should be conserved.

Fulfilling these criteria may lead to a conflict of goals since maximizing the effective population size suggests collecting extreme genotypes, which may not be representative for the population.

Using parameters from population genetics, the group of animals chosen should have minimum inbreeding and a minimum relationship to each other. Note that the level of inbreeding is less critical than the average relationship, because the actual inbreeding is removable while the level of relationship inevitably determines the long-term level of inbreeding.

If reliable pedigree data are available, these parameters can be calculated for any possible sample and used to identify the optimum group of animals to be stored. If pedigree information is not available, genetic markers can be used to approximate these criteria. Eding and Meuwissen (2001) suggested estimating Malécot's (1948) kinship coefficient based on molecular marker information and deriving what they call a "core set" to conserve.

7.4 CONCLUSIONS

Molecular markers are an indispensable tool to understand the genetic structures of populations. For the sampling of germplasm to create an animal gene bank, they are necessary but in no way sufficient to make adequate decisions. In addition to diversity information derived from molecular data, there needs to be good, specific knowledge and understanding of breed characteristics and values, the risk status of breeds, and availability and cost efficiency of possible conservation programmes, among others. Considering the present situation in livestock, diversity information is often more easily accessible than information on many of the other factors listed (Ruane, 2000). It is therefore strongly recommended to concentrate co-ordinated genotyping efforts to fill in the still existing "white spots" on the global maps of farm animal diversity and to re-allocate funds to develop a better understanding of the other components of a rational decision-making process.

7.5 REFERENCES

Baumung, R., Simianer, H. & Hoffmann, I. 2004. Genetic diversity studies in farm animals - a survey. *J. Anim. Breed. Genet.*, 121: 361-373.

Caballero, A. & Toro, M.A. 2002. Analysis of genetic diversity for the management of conserved subdivided populations. *Conservation Genetics*, 3: 289-299.

Coltman, D.W., Wilson, K., Pilkington, J.G., Stear, M.J. & Pemberton, J.M. 2001. A microsatellite polymorphism in the gamma interferon gene is associated with resistance to gastrointestinal nematodes in a naturally-parasitized population of Soay sheep. *Parasitology*, 122: 571-582.

Convention on Biological Diversity. 1992. Secretariat of the Convention on Biological Diversity. Montreal, Canada. (available at www.biodiv.org)

Eding, H. & Meuwissen, T.H.E. 2001. Marker based estimates of between and within population kinships for the conservation of genetic diversity. *J. Anim. Breed. Genet.*, 118: 141-159.

FAO. 2000. *World watch list for domestic animal diversity.* 3rd edition. (also available at dad.fao.org/en/refer/library/wwl/wwl3.pdf).

Hall, S.J.G. 2004. Livestock biodiversity. Genetic resources for the farming of the future. Oxford, England, Blackwell.

Hanotte, O., Ronin, Y., Agaba, M., Nilsson, P., Gelhaus, A., Horstmann, R., Sugimoto, Y., Kemp, S., Gibson, J., Korol, A., Soller, M. & Teale, A. 2003. Mapping of quantitative trait loci controlling trypanotolerance in a cross of tolerant West African N'Dama and susceptible East African Boran cattle. *Proc. Natl. Acad. Sci. USA*, 100: 7443-7448.

Hoffmann, I., Ajmone Marsan, P., Barker, J.S.F., Cothran, E.G., Hanotte, O., Lenstra, J.A., Milan, D., Weigend, S. & Simianer, H. 2004. New MoDAD marker sets to be used in diversity studies for the major farm animal species: recommendations of a joint ISAG/FAO working group. *Proceedings of the 29th International Conference on Animal Genetics*, Tokyo, Japan, 11-16 September, 2004: 123. (abstract)

Kimura, M. & Crow, J.F. 1963. On the maximum avoidance of inbreeding. *Genet. Res. Camb.*, 4: 399-415.

Malécot, G. 1948. *Les mathématiques de l'hérédité.* Paris, France, Masson et Cie. 64 pp.

Meijerink, E., Neuenschwander, S., Fries, R., Dinter, A., Bertschinger, H.U., Stranzinger, G. & Vögeli, P. 2000. A DNA polymorphism influencing alpha (1,2) fucosyltransferase activity of the pig FUT1 enzyme determines susceptibility of small intestinal epithelium to Escherichia coli F18 adhesion. *Immunogenetics*, 52: 129-136.

Pinent, T., Simianer, H. & Weigend, S. 2005. Weitzman's approach and conservation of breed diversity: First application to German chicken breeds. In *Proceedings of the International Workshop on the Role of Biotechnology for the Characterisation and Conservation of Crop, Forestry, Animal and Fishery Genetic Resources.* (available at www.fao.org/biotech/docs/pinent.pdf).

Piyasatian, N. & Kinghorn, B.P. 2003. Balancing genetic diversity, genetic merit and population viability in conservation programmes. *J. Anim. Breed. Genet.*, 120: 137-149.

Reist-Marti, S.B. 2004. Analysis of methods for efficient biodiversity conservation with focus on African cattle breeds. Zurich, Switzerland, Swiss Federal Institute of Technology. (PhD thesis)

Ruane, J. 2000. A framework for prioritizing domestic animal breeds for conservation purposes at the national level: a Norwegian case study. *Conservation Biology*, 14: 1385-1393.

Signorello, G. & Pappalardo, G. 2003. Domestic animal biodiversity conservation: a case study of rural development plans in the European Union. *Ecol. Econ.*, 45: 487-499.

Simianer, H. 2002. Noah's dilemma: Which breeds to take aboard the ark? CD-ROM communication no. 26-02. *Proceedings of the 7th World Congress on Genetics Applied to Livestock Production* (WCGALP). Montpellier, France.

Simianer, H. 2005. Decision making in livestock conservation. *Ecol. Econ.* (in print)

Simianer, H., Marti, S.B., Gibson, J., Hanotte, O. & Rege, J.E.O. 2003. An approach to the optimal allocation of conservation funds to minimise loss of genetic diversity between livestock breeds. *Ecol. Econ.*, 45: 377-392.

Thaon d'Arnoldi, C., Foulley, J.L. & Ollivier, L. 1998. An overview of the Weitzman approach to diversity. *Genetics Selection Evolution*, 30:149-161.

Weitzman, M.L. 1992. On diversity. *Quarterly Journal of Economics,* CVII: 363-405.

Weitzman, M.L. 1993. What to preserve? An application of diversity theory to crane conservation. *Quarterly Journal of Economics,* CVIII: 157-183.

Wilmut, I., Schnieke, A.E., McWhir, J., Kind, A.J. & Campbell, K.H.S. 1997. Viable offspring derived from fetal and adult mammalian cells. *Nature*, 385: 810-813.

8. Genetic characterization of livestock populations and its use in conservation decision-making

Olivier Hanotte and Han Jianlin

8.1 SUMMARY

This chapter reviews the importance and putative impacts of molecular markers on decision-making for livestock genetic resource conservation. Livestock diversity is shrinking rapidly and there is an urgent need to define strategies to prioritize breed conservation. For most livestock species a large number of genetic markers that show different Mendelian patterns of inheritances (maternal, paternal, bi-parental) are now available. Applied at a large geographic scale to the study of livestock populations, they provide information on centres of origins and migration routes as well as identify geographic areas of admixture among populations of different genetic origins. Such information can guide the choice of breeds and geographic areas for conservation actions. Calculations within and between diversity parameters allow selection of priority breeds for conservation to maximize diversity conserved for the benefit of future human generations.

8.2 INTRODUCTION

The *World Watch List for Domestic Animal Diversity* (3rd ed.) documents more than 6 300 breeds of breeds of livestock belonging to 30 domesticated species (FAO, 2000). These breeds were developed following domestication and natural and human selection over the past 12 000 years. The current number of breeds is likely an underestimation since a large proportion of indigenous livestock populations of the developing world, where most animal genetic resources are found today, have yet to be described at phenotypic and genetic levels. Livestock populations have evolved a unique adaptation to their agricultural production system and agro-ecological environments. Their genetic diversity has provided the material for the very successful breeding improvement programmes of the developed world in the 19th and 20th century. This represents a unique resource to respond to the present and future needs of livestock production, both in developed and developing countries.

However, livestock diversity is shrinking rapidly. With the exception of the wild boar, which is the ancestor of the domestic pig, and wild red jungle fowl, which is the ancestor of the domestic chicken, the putative wild ancestors of our major livestock species, the reservoir of genetic diversity, are now either extinct (e.g. the auroch, the wild ancestor of cattle, or the ancestral species of the Old World camelids) or low in numbers and threatened by extinction (e.g. wild goat populations of the Near East, vicuña from Andean plateau, wild donkey in Africa). Among the domesticated populations, it is estimated that one to two breeds are lost every week (FAO, 2000). However, the impact of these losses on global or local diversity remains undocumented. While it is already too late for many breeds in Europe, the situation is also particularly worrying in the developing world where rapid changes in production systems are leading to the replacement of breeds or at best crossbreeding. There is therefore an urgent need to document the diversity of our livestock genetic resources and to design strategies for their sustainable conservation.

The task is enormous. It has prompted FAO and other international organizations to develop domestic animal diversity information systems and databases (ILRI's Domestic Animal Genetic Information System [DAGRIS] and FAO's DAD-IS 2.0). More recently, FAO has initiated a major country-driven documentation exercise, the *State of the World's Animal Genetic Resources* (SoW-AnGR). It is hoped that this book, together with the publication of a companion Strategic Priority Action Report, will lead to immediate actions for conservation on the ground at the country or regional level (SoW-AnGR, http://dad.fao.org/cgi-dad/$cgi_dad.dll/sow?eng).

To put this plan into practice, effective conservation of animal genetic resources (AnGR), whether *in situ* or *ex situ*, will require the mobilization of substantial economic resources over a long period of time. Financial resources are limited and they always will be. Methods are required to identify priority decisions in order to maximize the diversity conserved, both at the local and the global level, and to focus on unique genetic resources of global significance. Genetic characterization through the use of molecular markers associated to powerful statistical approaches is providing new avenues for decision-making choices for the conservation and rational management of AnGR.

8.3 GENETIC CHARACTERIZATION TOOLS: MOLECULAR MARKERS

Protein polymorphisms were the first molecular markers used in livestock. Many studies, particularly during the 1970s, have documented the characterization of blood group and allozyme systems of livestock (e.g. Baker and Manwell, 1980). However, the level of polymorphism observed in proteins is often low, which has reduced the general applicability of protein typing in diversity studies. With the development of the Polymerase Chain Reaction (PCR) and sequencing technologies associated with automatic and/or semi-automatic large-scale screening systems, DNA-based polymorphisms are now the markers of choice for molecular-based surveys of genetic variation. Importantly, polymorphic DNA markers showing different patterns of Mendelian inheritances can now be studied in nearly all of humankind's major livestock species. Typically, they include D-loop and cytochrome B mitochondrial DNA (mtDNA) sequences (maternal inheritance), Y chromosome-specific single nucleotide polymorphisms (SNPs) and microsatellites (paternal inheritance) and autosomal microsatellites bi-parental inheritance) (Avise, 1994). Interestingly, while recent developments in cytogenetic technologies should facilitate the isolation of Y chromosomes specific markers (Petit, Balloux and Excoffier, 2002), for most livestock species there are still few Y polymorphic markers. This is possibly a consequence of the demographic history of domestication and breed formation. In polygenous species, as in most livestock, we clearly expect that a small number of male lineages would have contributed to the genetic pool of the species. Polymorphic Y microsatellite markers are currently only available for cattle (Hanotte *et al.*, 2000), yak (Xuebin, 2004), and to some extent,. small ruminants (Lenstra and Econogene Consortium, 2005). To the best of the knowledge available, they have not yet been isolated in any major livestock species, e.g. the Old and New World camelids or the domestic pig. On the other hand, autosomal microsatellites have now been isolated in large numbers from most livestock species and recommended FAO/ISAG lists of autosomal microsatellite markers for genetic characterization studies are publicly available (FAO, 2004).

Important assumptions on the use of genetic markers include: (i) that the polymorphisms observed at the molecular markers are neutral; and (ii) that the use of a relatively small number of independently segregating marker loci will be a good predictor of the overall genomic diversity of a population, or in other words, that variation in allele frequencies between populations will reflect the distribution of genetic diversity within and among populations.

8.4 DIVERSITY OF GENETIC CHARACTERIZATION INFORMATION

Genetic markers will provide different types of information relevant for conservation-making decisions for livestock (Sunnucks, 2000). Autosomal microsatellite loci will be commonly used for individual genetic identification and parentage analysis, e.g. for the successful implementation and monitoring of *ex situ* conservation programmes, population diversity estimation, differentiation of populations, calculation of genetic distances, genetic relationships and population genetic admixture estimation. Microsatellite loci are also highly sensitive to genetic bottlenecks and are commonly used for inbreeding estimation. MtDNA sequences will be the markers of choice for domestication studies because the segregation of a mtDNA lineage within a livestock population will only have occurred through the domestication of a wild female or through the incorporation of a female into the domestic stock. More particularly, mtDNA sequences will be used to identify the putative wild progenitors, the number of maternal lineages and their geographic origins. To some extent, they may provide important information on the geographic distribution of diversity within livestock species although the usefulness of mtDNA sequences data will vary between species, depending on the demographic history of the migration from the centre(s) of domestication. Finally, the study of a diagnostic Y chromosome polymorphism is an easy and rapid way to detect and quantify male-mediated admixture.

A surprising result of the application of molecular makers in genetic characterization of livestock has been the discovery that several ancestral species, subspecies or maternal lineages have contributed to today's genetic pool of our major livestock species (Xuebin, 2004; Bruford, Bradley and Luikart, 2003; Beja-Pereira, 2004; ILRI unpublished data). It is clear from these recent results that multiple domestications and/or maternal introgression are the rule, not the exception (Table 11).

More particularly, mtDNA information supports the conclusion that there were at least five major centres of livestock domestication: the northern Andean chain (New World camelids), the Northeast African region (donkey and likely taurine cattle), the Near East (taurine cattle, sheep, goat, pigs), South Asia (Indus Valley, indicine cattle and chicken) and East Asia (pigs, chicken, horse, buffalo), to which the Hindu-Kush Himalayan region (yak) and North and Central Asia (horse) should be added. Similarly, to some extent we could expect multiple male-mediated introgression lineages. This is the case in the yak (Xuebin, 2004), with two distinct male lineages, but not the case in the domestic horse where screening for SNPs in 52 stallions from 15 different breeds did not identify a single polymorphic site (Lindgren *et al.* 2004).

Taking into account the history of human migration and trading, it is expected that our indigenous breeds today will often have multiple genetic signatures of

TABLE 11
No. and putative centres of origin of major maternal lineage in livestock

Domestic species	Number of maternal lineages	Geographic centres of origin
Cattle		
Bos taurus	2	Near East, Northeast Africa
Bos indicus	1	South Asia (Indus Valley)
Yak		
Bos grunnies	3	Hindu-Kush Himalayan region
Sheep		
Ovis aries	3	Near East, South Asia
Goat		
Capra hircus	At least 3	Near East, South Asia
Horse		
Equus caballus	Multiple	North and Central Asia
Donkey		
Equus asinus	2	Northeast Africa
Pig		
Sus scrofa domesticus	2	Near East, East Asia
Water Buffalo		
Bubalus bubalis bubalis	1	South Asia
Bubalus bubalis carabensis	1	East Asia
Llama		
Llama glama	4	Northern Andean chain
Alpaca		
Vicugna pacos	4	Northern Andean chain

Source: Bruford *et al.* (2003), Xuebin (2004) [yak], Beja-Pereira (2004) [donkey].

origin and admixture. Available molecular data indicate that ancient genetic admixtures between livestock populations from different domestication events are common on the Asian continent and, to some extent, also in Africa, but mostly absent from Europe. This has been shown in cattle (e.g. Hanotte *et al.*, 2002), pigs and small ruminants (Bruford, Bradley and Luikart, 2003).

8.5 GENETIC CHARACTERIZATION AND CONSERVATION DECISION-MAKING

As illustrated above, genetic characterization is providing new information to guide and prioritize conservation decisions for livestock. Possibly, the most urgently required action is the effective protection of all remaining wild ancestral populations and closely related species of livestock, most of them now endangered. They are the only remaining sources of putative alleles of economic values that might have been lost during domestication events. Coordination with international wildlife conservation institutions such as the World Conservation Union (IUCN) is required. It is equally important to ensure that the breeds selected for conservation include populations from the geographic areas representing the different domestication centres where we would expect to find large genetic diversity and genetically differentiated populations. Animals and populations present at the geographic area of a centre of domestication will also be expected to be very distinct from the ones found at other centres of domestication. Also, understanding the geographic pattern of livestock migration from a centre of origin will allow the identification of populations present at the end of a migration route. It is expected that these populations will be genetically distinct from the populations present at the ends of other migration routes as a result of random genetic drift and/or the effect of local selection pressures. Importantly, knowledge of both the global diversity of the breeds and admixture events will be needed in order to be able to make sound priority decisions.

The next challenge is to make priority decisions for conservation among today's thousands of domestic breeds or populations. The primary objective is to maximize the conservation of the genetic diversity available for potential future use. At the ideal extreme, this would be achieved through the conservation of all breeds of livestock. Such a comprehensive approach would ensure complete conservation of diversity. In practice, it is unrealistic and prioritization of actions will have to be made. Two criteria (perhaps to be eventually combined) have been proposed (Gibson, Ayalew, and Hanotte, in press): priority breeds for conservation should be those with the largest within-breed diversity and/or should maximize the conservation of between breed diversity. Both within and between breed diversity parameters are classically measured using molecular genetic markers. In both cases, soundly based priority decisions for conservation at the global level will require the availability of large datasets.

The mean number of alleles (MNA) and observed (*Ho*) and expected (*He*) heterozygosity are the most commonly calculated population genetic parameters for assessing within-breed diversity. For example, in a recent study, three distinct sets of microsatellite diversity cattle data were merged to provide, for the first time, within breed diversity (*He* and MNA), as well as admixture information combined for Europe, Africa, the Near East and South Asia (Freeman *et al.*, in preparation). The geographic region with the highest diversity is found between the two likely Asian centres of cattle domestication, in a broad geographic area corresponding to what are today Iran, Iraq and the Caucasian region. Global geographic analysis of admixture suggests that the region corresponds to a geographic area of around 50 percent admixture between taurine and indicine cattle. Genetic diversity and admixture information from more indigenous breeds are needed to confirm the results. If confirmed, this geographic area will undoubtedly represent a major livestock diversity hotspot, a priority region for a global plan for the conservation of the diversity of domestic cattle.

The simplest parameters for assessing the distribution of diversity between breeds using genetic markers are the genetic differentiation or fixation indices (e.g. F_{ST}, G_{ST}, θ). The most widely used is F_{ST}, which measures the degree of genetic variation between subpopulations through the calculation of the standardized variances in allele frequencies among populations (Weir and Basten, 1990, Mburu *et al.*, 2003). Any set of genetic distances can also be analysed in terms of between-breed genetic diversity, and more particularly, in terms of individual breed contributions to the total diversity of a set of breeds. The most common approach used so far is a method proposed by Weitzman (1993). It involves calculating a matrix of genetic distances and generating dendrograms. Individual breed contributions are calculated by comparing the total length of the dendrogram including all breeds with the dendrogram including all breeds less the individual breed. Priority breeds for conservation would therefore be the breeds contributing most to the diversity of the set. The method can be extended further to estimate the impact of conservation decisions on the diversity of a set of breeds in the future with the calculation of the extinction probability of each breed and the marginal diversity that reflect the relative loss or gain in expected diversity of a set of breeds following a decrease or increase in the probability of survival of a breed by one unit.

The largest data set to which the Weitzman approach has been applied in livestock is on 49 African cattle breeds (Reist-Marti *et al.*, 2003). The breeds were divided into two groups corresponding to the "taurine" and "indicine" division, and extinction probabilities were calculated for each breed. In both groups the results clearly indicate that the optimal conservation strategy is to give priority to the breeds with the highest marginal diversity rather than the most endangered ones.

8.6 FUTURE CHALLENGES AND OPPORTUNITIES

Major challenges remain for livestock conservationists. Documentation of genetic diversity is still all but lacking for some livestock species and incomplete for others (e.g. Old World camelids, chicken, buffalo, Asian small ruminants and cattle, etc.). With a few exceptions (see, for example, Hanotte *et al.*, 2002), molecular datasets will include in a single study a limited number of breeds or populations only, and combined analysis of molecular datasets obtained in different studies will often be impossible. For example, statistical approaches are still lacking that allow the combination of genetic-distancing information obtained in separate studies. Finally, the molecular markers used to characterize diversity have little to do with the genes under selection for economically important traits (Gibson, Ayalew and Hanotte, in press).

But there are promising and exciting new avenues. We can expect that with the increased adoption of common sets of markers and common breeds of references (Freeman *et al.*, in preparation), the combination of microsatellite datasets will be facilitated. The publication of the entire genome sequences of several livestock species will allow for the easy identification of thousands of neutral and selected genetic markers. It will open the way to detecting signatures of selection allowing researchers to trace the presence and the spread of economically important alleles (Luikart *et al.*, 2003). A recent study in cattle milk protein genes has indirectly yet well illustrated the putative application of such selected markers in the identification of breeds and geographic areas as priorities for the conservation of specific economically important traits (Beja-Pereira *et al.*, 2003). The new field of livestock "landscape genetics" is emerging (Manel, 2003). It will combine geo-referencing of breed distributions, spatial global genetic diversity, and climatic, ecological, epidemiological and production system information, which will facilitate and direct priority decisions for *in situ* breed conservation.

8.7 REFERENCES

Avise, J.C. 1994. *Molecular markers, natural history and evolution.* New York, USA, Chapman and Hall.

Baker, C.M.A & Manwell, C. 1980. Chemical classification of cattle. I. Breed group. *Animal Blood Groups and Biochemical Genetics,* 11: 127-150.

Beja-Pereira, A. 2004. African origins of the domestic donkey. *Science,* 304: 1781.

Beja-Pereira, A., Luikart, G., England, P.R., Bradley, D.G., Jann, O.C., Bertorelle, G., Chamberlain, A.T., Nunes, T.P., Metodiev, S., Ferrand, N. & Erhardt, G. 2003. Gene-culture coevolution between cattle milk protein genes and human lactase genes. *Nature Genetics,* 35: 311-313.

Bruford, M.W., Bradley, D.G. & Luikart, G. 2003. DNA markers reveal the complexity of livestock domestication. *Nature Review Genetics,* 4: 900-910.

FAO. 2000. *World watch list for domestic animal diversity.* 3rd edition. Rome. (also available at dad.fao.org/en/refer/library/wwl/wwl3.pdf).

FAO. 2004. Secondary guidelines for development of national farm animal genetic resources management plans. Measurement of domestic animal diversity (MoDAD): recommended microsatellite markers. (available at dad.fao.org/en/refer/library/guidelin/marker.pdf).

Freeman, A.R., Bradley, D.G., Nagda, S., Gibson, J.P. & Hanotte, O. (in preparation). *Combination of multiple microsatellite datasets to investigate genetic diversity and admixture.*

Gibson, J.P., Ayalew, W. & Hanotte, O. (in press). *Measures of diversity as inputs for decisions in conservation of livestock genetic resources.* Wallingford, Oxfordshire, England, CABI.

Hanotte, O., Bradley, D.G., Ochieng, J.W., Verjee, Y., Hill, E.W. & Rege, J.E. 2002. African pastoralism: genetic imprints of origins and migrations. *Science*, 296: 336-339.

Hanotte, O., Tawah, C.L., Bradley, D.G., Okomo, M., Verjee, Y., Ochieng, J. & Rege, J.E. 2000. Geographic distribution and frequency of a taurine *Bos taurus* and an indicine *Bos indicus Y* specific allele amongst sub-Saharan Africa cattle breeds. *Molecular Ecology*, 9: 387-396.

Lenstra, J.A. & Econogene Consortium. 2005. Evolutionary and demographic history of sheep and goats suggested by nuclear, mtDNA and Y-chromosome markers. In *Proceedings of the International Workshop on the role of biotechnology for the characterisation and conservation of crop, forestry, animal and fishery genetic resources.* (available at www.fao.org/biotech/docs/lenstra.pdf).

Lindgren, G., Backström, N., Swinburne, J., Hellborg, L., Einarsson, A., Sandberg, K., Vilà, C., Binns, M. & Ellegren, H. 2004. Limited number of patrilines in horse domestication. *Nature Genetics*, 36: 335-336.

Luikart, G., England, P.R., Tallmon, D., Jordan, S. & Taberlet, P. 2003. The power and promise of population genomics: from genotyping to genome typing. *Nature Review Genetics*, 4: 981-994.

Manel, S. 2003. Landscape genetics: combining landscape ecology and population genetics. *Trends in Ecology and Evolution*, 18: 189-197.

Mburu, D.N., Ochieng, J.W., Kuria, S.G., Jianlin, H., Kaufmann, B., Rege, J.E. & Hanotte, O. 2003. Genetic diversity and relationships of indigenous Kenyan camel *(Camelus dromedarius)* populations: implications for their classification. *Animal Genetics*, 34: 26-32.

Petit E., Balloux F. & Excoffier L. 2002. Mammalian population genetics: Why not Y? *Trends in Ecology and Evolution*, 17: 28-33.

Reist-Marti, S.B., Simianer, H., Gibson, J., Hanotte, O. & Rege, J.E.O. 2003. Weitzman's approach and conservation of breed diversity: an application to African cattle breeds. *Conservation Biology*, 17: 1299-1311.

Sunnucks, P. 2000. Efficient genetic markers from population biology. *Trends in Ecology & Evolution*, 15: 199-203.

Weir, B.S. & Basten, C.J. 1990. Sampling strategies for distances between DNA sequences. *Biometrics*, 46: 551-582.

Weitzman, M.L. 1993. What to preserve? An application of diversity theory to crane conservation. *Quarterly Journal of Economics*, 108: 157-183.

Xuebin, Q. 2004. Genetic diversity, differentiation and relationship of domestic yak populations: a microsatellite and mitochondrial DNA study. University of Berne. Berne, Switzerland, (PhD thesis). 62 pp.

9. Genetic characterization of populations and its use in conservation decision-making in fish

Craig Primmer

9.1 SUMMARY

The conservation needs of fishes are special in many respects compared to those of terrestrial organisms. For example, aquatic organisms are the only major human food source that is primarily harvested from wild populations. This chapter will briefly outline some of the main applications in which molecular marker data are applied for conservation decision-making in fish populations, which include determining population genetic structure, estimation of effective population size and detection of population size changes.

The needs for fish conservation are special in many respects compared to those of terrestrial organisms and in particular, domestic species that have been considered in other chapters of this publication. For example, aquatic organisms are the only major human food source that are primarily harvested from wild populations (Ryman *et al.*, 1995). Therefore, methods for the development of population management guidelines often more closely follow those commonly used for wildlife than for domestic animals and plants. This chapter will briefly outline some of the main applications for which molecular marker data are applied for preserving fish biodiversity.

9.2 DETERMINING POPULATION GENETIC STRUCTURE

One of the main applications of molecular markers related to the conservation of commercially exploited fish is to aid in the development of guidelines enabling sustainable harvesting of populations (Carvalho and Pitcher, 1994). An important first step towards this is to characterize the genetic structure of the populations being harvested, which will assist in defining the biological units that should be considered when developing a management strategy, the so-called "genetic stock concept" (e.g. Carvalho and Hauser, 1994). While not directly aimed at conserving biodiversity, but rather at maximizing sustainable harvest levels, the implementation of management strategies based on molecular data can have indirect benefits for population biodiversity since their aim is to avoid population crashes that would negatively affect harvesting. This aim also benefits the maintenance of population genetic diversity. Studies investigating the population genetic structure of fishes have been conducted for a number of decades, first using protein coding allozyme loci (e.g. Utter, Aebersold and Winans, 1987) and then starting in the mid-1980s, using mitochondrial DNA (mtDNA) polymorphisms (reviewed by Avise, 2004). More recently, tandemly repeated microsatellite DNA markers have become the molecular marker of choice for determining intraspecific population genetic relationships (for example, Koskinen *et al.*, 2002a). In general, markers that are used today utilize the polymerase chain reaction (PCR) since they enable the analysis of archive material such as scales. (For more details regarding the different marker types, refer to Frankham, Ballou and Briscoe, 2002; Avise, 2004.)

From a fisheries management perspective, the aim of determining the intraspecific population genetic structure is to determine the units between which limited gene flow occurs: if such units are overfished, it is unlikely that population sizes will recover due to migration and hence a collapse of the fishery may occur. While understanding population genetic structure is important from an applied perspective, the same knowledge is also the basis of any biologically sound conservation strategy. For example, similar genetic criteria to those described above are the basis of several definitions of the so-called "Evolutionary Significant Unit" (ESU) (reviewed by Fraser and Bernatchez, 2001). For example, since 1978, in the United States, the Endangered Species Act (ESA) affords protection to "*any subspecies of fish or wildlife or plants, and any distinct population segment of any species of vertebrate fish or wildlife which interbreeds when mature*" (Endangered Species Act, Sec. 3 [15]). A distinct population segment (DPS) is not well described in the ESA, but some definitions have been developed. For example, the United States National Marine Fisheries Service (NMFS) developed two criteria for salmonid populations to be considered a DPS: it must be substantially reproductively isolated from other con-specific population units and it must represent an important component in the evolutionary legacy of the species.

Molecular marker information can therefore assist in defining ESUs, which in turn can be used for developing conservation strategies of populations. A related technique that has been useful for delineating the population of origin of individuals in mixed stock fisheries is individual assignment, where individuals are assigned to the population from which their multi-locus microsatellite genotype has the highest probability of occurring (see, for example, Cornuet *et al.*, 1999 and the program "GeneClass", available at www.montpellier.inra.fr/URLB). From a conservation perspective, assignment tests can be useful for distinguishing native born individuals from those that are hatchery-reared, or for identifying individuals whose genetic make-up most closely resembles the original stock, based on, for example, archived scale data (see the review in Hansen, Kenchington and Nielsen, 2001).

9.3 UNDERSTANDING THE GENERAL EFFECTS OF DIFFERENT AQUATIC ENVIRONMENTS ON FISH GENETIC DIVERSITY AND DIFFERENTIATION

Based on the wealth of population genetic research already conducted in fish, several studies summarizing data based on allozymes (Gyllensten 1985; Ward, Woodwark and Skibinski, 1994) and microsatellites (DeWoody and Avise, 2000) have been published that demonstrate some general trends in genetic structuring for species occupying different habitats during their life cycle. These studies have revealed marked differences in the level of genetic differentiation and genetic diversity between populations of marine and freshwater species, with marine species generally exhibiting lower levels of inter-population differentiation (Gyllensten, 1985; Ward, Woodwark and Skibinski, 1994) and higher genetic diversity (Gyllensten, 1985; Ward, Woodwark and Skibinski, 1994; DeWoody and Avise, 2000). This general observation has generally been hypothesized as a result of higher effective population sizes and/or higher inter-population migration rates in marine than in freshwater environments and has implications for the conservation of genetic diversity. Lower effective population sizes and/or lower inter-population migration rates in the freshwater environment predict that populations of freshwater species will be more prone to extinction than marine species and thus worthy of particular conservation concern. This does not imply, however, that marine populations are immune to such effects (see below).

9.4 ESTIMATION OF EFFECTIVE POPULATION SIZE

Accurate estimates of effective population size (*Ne*) are central to the development of appropriate conservation strategies in any species because *Ne* predicts the rate of loss of neutral genetic variation, the fixation rate of deleterious and favourable alleles, and the rate of increase of inbreeding experienced by a population, etc. (Frankham, Ballou and Briscoe, 2002). Importantly, the *Ne* is often many times smaller than the census size (*N*) of the population; the *Ne/N* ratio averages just 0.11 in a survey of vertebrate species (Frankham, 1995). In fish, *Ne/N* ratios may be expected to be even more extreme due to the high female fecundity of many species enabling large census numbers to be obtained from minimal numbers of breeding individuals. For example, the winter chinook salmon run in the Sacramento River of California consists of around 2 000 mature individuals but the effective size of the population has been estimated to be only 85 (*Ne/N* = 0.04: Bartley *et al.*, 1992).

While estimates of *Ne* can be gained using direct methods based on field data (estimates of sex ratio bias, offspring production, variation in family size, etc.), obtaining such data can be very cumbersome in many wild populations, especially in aquatic species. Hence, indirect methods for *Ne* estimation based on molecular marker data have also been developed. From a practical viewpoint, these methods can be broken down into two categories: those that require data from a single population sample (single generation methods, for example, Hill 1981; Pudovkin, Zaykin and Hedgecock, 1996; Beaumont, 1999; Luikart and Cornuet, 1999) and those requiring samples from the same population collected at least one generation apart (temporal methods: Waples, 1989; Anderson, Williamson and Thompson, 2000; Wang, 2001; Berthier *et al.*, 2002). The temporal methods generally utilize variation in temporal allele frequencies to estimate the level of genetic drift and hence the effective population size. This group of methods tends to give more reliable results than the single generation methods. An important recent advance has been the development of methods that take into consideration the effects of migration on *Ne* estimation (Wang and Whitlock 2003; the MLNE program is available from www.zoo.cam.ac.uk/ioz/software.htm). The major limitation in the use of these methods is that double the sampling effort is required. This can be a particularly big problem for late breeding species since collection of samples at least one generation apart can fall outside the time frame of a funded study. Nevertheless, due to the existence of historical scale samples in a number of commercially important species from which sufficient DNA can be extracted, originally collected for understanding the age structure of populations, temporal methods for *Ne* estimation have been applied relatively frequently in fish. In a high profile example, Hauser *et al.* (2002) demonstrated through microsatellite analyses of a time series of historical scale samples that New Zealand snapper (*Pagrus auratus*) has undergone a decline in genetic diversity during their exploitation history. In addition, effective population sizes were estimated at five orders of magnitude lower than estimated census sizes. This study provided one of the first indications that genetic problems can potentially exist in marine species for which individual numbers are in their millions.

9.5 DETECTION OF POPULATION SIZE CHANGES

From a conservation perspective, detection of recent dramatic changes in population size (population bottlenecks) is another important aspect of any population monitoring programme (Frankham, Ballou and Briscoe, 2002). Signals of past population bottlenecks can be detected using molecular genetic analyses. For example, one commonly applied method makes use of the assumption that populations that have experienced a recent reduction in *Ne* will show a reduction in both heterozygosity and allele number at polymorphic loci. However, the reduction in allele number is faster than the reduction in heterozygosity. Therefore, in a recently bottlenecked population, the observed heterozygosity is higher than the expected equilibrium heterozygosity when calculated from the observed number of alleles, under the assumption of a constant-size population (Luikart and Cornuet 1997). Several statistical tests have been developed to determine whether a population exhibits a significant number of loci with heterozygosity excess, hence indicating the occurrence of a population bottleneck (Cornuet and Luikart 1996; implemented in the program *Bottleneck* available at www.montpellier.inra.fr/URLB). Additional tests for identifying reductions or expansions in population size based on DNA sequence data (Rogers and Harpending, 1992; Templeton, 1998) or allele frequency data (e.g. Beaumont, 1999) have also been developed. The choice of method depends on a number of factors including the samples available, the molecular marker data available and the time frame of any potential bottlenecks.

An important feature of all the molecular data applications described above is that temporal information i.e. data collected for the same populations over an extended time period, can be extremely valuable. Fortunately, due to the existence of historical scale samples in a number of commercially important species, the collection of such data is often more feasible in fish than in other taxonomic groups. This has been utilized to good effect by fish conservation geneticists (e.g. Nielsen, Hansen and Loeschcke, 1997; Nielsen, Hansen and Loeschcke, 1999; Hansen *et al.*, 2002, Hansen *et al.*, 2002; Koskinen *et al.*, 2002b). In addition, regular monitoring of populations is important for enabling a distinction between normal population size fluctuations and those severe enough to warrant conservation measures (Laikre, 1999).

9.6 PRIORITIZING POPULATIONS FOR CONSERVATION

Given that resources for preserving biodiversity are limited, there is an important need for biologically sensible criteria that can be applied simply to prioritize fish genetic diversity conservation efforts. Currently, however, such methods are rarely applied (for an exception, however, see Allendorf *et al.*, 1997), although decisions are often made in a rather *ad hoc* manner or based on solely non-biological criteria such as monetary value. While such criteria should not necessarily be ignored, they should not completely replace biological criteria. Although methods for identifying populations harbouring a higher proportion of a species genetic diversity exist (e.g. Crozier, 1997; Caballero and Toro, 2002; Reist-Marti *et al.*, 2003; Simianer *et al.*, 2003), such methods are yet to be applied in a fish biodiversity preservation context. One model outlining a method for population conservation prioritization, based on experiences in brown trout is outlined in Laikre (1999). This model combines molecular genetic, phenotypic and socio-economic/cultural criteria to prioritize populations for conservation.

9.7 CONFLICTING NEEDS OF DIFFERENT SECTORS OF SOCIETY

As noted above, due to the commercial and cultural importance (e.g. recreational angling) of many fish populations, conservation guidelines of commercially and/or culturally important populations potentially conflict with the needs of other interest groups such as commercial and recreational anglers. Therefore, a challenge when developing a conservation programme for such populations is to find a balance that satisfies all groups, including researchers, decision-makers and end-users. While considerable effort will undoubtedly be put into improving the analytical methodologies described above, it should also be recognized that efforts aimed at decreasing this imbalance are likely to be equally important for fish biodiversity preservation and increased communication between these groups should be encouraged.

9.8 REFERENCES

Allendorf, F.W., Bayles, D., Bottom, D.L., Currens, K.P., Frissell, C.A., Hankin, D., Lichatowich, J.A., Nehlsen, W., Trotter, P.C. & Williams, T.H. 1997. Prioritizing Pacific salmon stocks for conservation. *Conservation Biology,* 11:140-152.

Anderson, E.C., Williamson, E.G. & Thompson, E.A. 2000. Monte Carlo evaluation of the likelihood for Ne from temporally spaced samples. *Genetics*, 156: 2109-2118.

Avise, J.C. 2004. *Molecular markers, natural history and evolution,* 2nd edition. New York, USA, Chapman & Hall. 511 pp.

Bartley, D., Bagley, M., Gall, G. & Bentley, B. 1992 Use of linkage disequilibrium data to estimate effective size of hatchery and natural fish populations. *Conservation Biology,* 6: 365-375.

Beaumont, M. 1999. Detecting population expansion and decline using microsatellites. *Genetics,* 153: 2013-2029.

Berthier, P., Beaumont, M.A., Cornuet, J-M. & Luikart, G. 2002 Likelihood-based estimation of the effective population size using temporal changes in allele frequencies: a genealogical approach. *Genetics,* 160: 741-751.

Caballero, A. & Toro, M.A. 2002. Analysis of genetic diversity for the management of conserved subdivided populations. *Conservation Genetics,* 3: 289-299.

Carvalho, G.R. & Hauser, L. 1994. Molecular-genetics and the stock concept in fisheries. *Reviews in Fish Biology and Fisheries,* 4: 326-350.

Carvalho, G.R. & Pitcher, T.J. 1994. *Molecular genetics in fisheries.* London, UK, Chapman & Hall. 131 pp.

Cornuet, J.M. & Luikart, G. 1996. Description and power analysis of two tests for detecting recent population bottlenecks from allele frequency data. *Genetics,* 144: 2001-2014.

Cornuet, J., Piry, S., Luikart, G., Estoup, A. & Solignac, M. 1999. New methods employing multilocus genotypes to select or exclude populations as origins of individuals. *Genetics,* 153: 1989-2000.

Crozier, R.H. 1997. Preserving the information content of species: genetic diversity, phylogeny, and conservation worth. *Annual Reviews in Ecology and Systematics,* 28: 243-268.

DeWoody, J.A. & Avise, J.C. 2000. Microsatellite variation in marine, freshwater and anadromous fishes compared with other animals. *Journal of Fish Biology,* 56: 461-473.

Frankham, R. 1995. Effective population size/adult population size ratios in wildlife: a review. *Genetical Research,* 66: 95-107.

Frankham, R., Ballou, J.D. & Briscoe, D.A. 2002. *Introduction to conservation genetics.* Cambridge, UK, Cambridge University Press.

Fraser, D.J. & Bernatchez, L. 2001. Adaptive evolutionary conservation: towards a unified concept for defining conservation units. *Molecular Ecology,* 10: 2741-2752.

Gyllensten, U. 1985. The genetic structure of fish: differences in the intraspecific distribution of biochemical genetic variation between marine, anadromous, and freshwater species. *Journal of Fish Biology,* 26: 691-699.

Hansen, M.M., Kenchington, E. & Nielsen, E.E. 2001. Assigning individual fish to populations using microsatellite DNA markers. *Fish & Fisheries,* 2: 93-112.

Hansen, M.M., Ruzzante, D.E, Nielsen, E.E., Bekkevold, D. & Mensberg, K.L.D. 2002. Long-term effective population sizes, temporal stability of genetic composition and potential for local adaptation in anadromous brown trout *(Salmo trutta)* populations. *Molecular Ecology,* 11: 2523-2535.

Hauser L, Adcock, G., Smith, P.J., Bernal Ramirirez, J.H. & Carvalho, G.R. 2002. Loss of microsatellite diversity and low effective population size in an overexploited population of New Zealand snapper *(Pagrus auratus). PNAS*, 99: 11742-11747.

Hill, W.G. 1981. Estimation of effective population size from data on linkage disequilibrium. *Genetical Research*, 38: 209-216.

Koskinen, M.T., Piironen, J., Sundell, P. & Primmer, C.R. 2002a. Spatiotemporal evolutionary relationships and genetic assessment of stocking effects in grayling (*Thymallus thymallus,* Salmonidae). *Ecology Letters,* 5: 193-205.

Koskinen, M.T., Ranta, E., Piironen, J., Veselov, A., Nilsson, J. & Primmer, C.R. 2002b. Microsatellite data detect low levels of intrapopulation genetic diversity and resolve phylogeographic patterns in European grayling, *Thymallus thymallus,* Salmonidae. *Heredity*, 88: 391-401.

Laikre, L. 1999. *Conservation genetic management of brown trout (Salmo trutta) in Europe. Report by the concerted action on identification, management and exploitation of genetic resources in the brown trout (Salmo trutta).* Division of Population Genetics, Stockholm University. 91 pp. (available at www.dfu.min.dk/ffi/consreport/index.htm).

Luikart, G. & Cornuet, J-M. 1997. Empirical evaluation of a test for identifying recently bottlenecked populations from allele frequency data. *Conservation Biology,* 12: 228-237.

Luikart, G. & Cornuet, J-M. 1999. Estimating the effective number of breeders from heterozygote excess in progeny. *Genetics,* 151: 1211-1216.

Nielsen, E.E., Hansen, M.M. & Loeschcke, V. 1997. Analysis of microsatellite DNA from old scale samples of Atlantic salmon: a comparison of genetic composition over sixty years. *Molecular Ecology,* 6: 487-492.

Nielsen, E.E., Hansen, M.M. & Loeschcke, V. 1999. Genetic variation in time and space: microsatellite analysis of extinct and extant populations of Atlantic salmon. *Evolution,* 53: 261-268.

Pudovkin, A.I., Zaykin, D.V. & Hedgecock, D. 1996. On the potential for estimating the effective number of breeders from heterozygote excess in progeny. *Genetics,* 144: 383-387.

Reist-Marti, S.B., Simianer, H., Gibson, J., Hanotte, O. & Rege, J.E.O. 2003. An approach to the optimal allocation of conservation funds to minimize losses of genetic diversity between livestock breeds. *Conservation Biology,* 17: 1299-1311.

Rogers, A.R. & Harpending, H. 1992. Population growth makes waves in the distribution of pairwise genetic differences. *Molecular Biology and Evolution,* 9: 552-569.

Ryman, N., Utter, F. & Laikre L. 1995. Protection of intraspecific biodiversity of exploited fishes. *Reviews in Fish Biology and Fisheries,* 5: 417-446.

Simianer H., Marti S.B., Gibson, J., Hanotte, O. & Rege, J.E.O. 2003. An approach to the optimal allocation of conservation funds to minimize losses of genetic diversity between livestock breeds. *Ecological Economics,* 45: 377-392.

Templeton, A.R. 1998. Nested clade analysis of phylogeographic data: testing hypotheses about gene flow and population history. *Molecular Ecology,* 7: 381-398.

Utter, F., Aebersold, P. & Winans, G. 1987. Interpreting genetic variation detected by electrophoresis. *In* Ryman, N. & Utter, F., eds. *Population genetics and fishery management.* Seattle, Washington, USA, Washington Sea Grant Program/University of Washington Press.

Wang, J. 2001. A pseudo-likelihood method for estimating effective population size from temporally spaced samples. *Genetical Research,* 78: 243-257.

Wang, J. & Whitlock, M.C. 2003. Estimating effective population size and migration rates from genetic samples over space and time. *Genetics,* 163: 429-446.

Waples, R.S. 1989. A generalised approach for estimating effective population size from temporal changes in allele frequency. *Genetics,* 121: 379-391.

Ward, R.D., Woodwark, M. & Skibinski, D.O.F. 1994. A comparison of genetic diversity levels in marine, freshwater, and anadromous fishes. *Journal of Fish Biology,* 44: 213-232.

10. Molecular marker based analysis for crop germplasm preservation

Sergio Lanteri and Gianni Barcaccia

10.1 SUMMARY

This chapter aims to highlight the potential impact and possible drawbacks of the application of DNA marker technology to crop germplasm preservation. Some of the issues related to crop management and use, which can be addressed through information derived from DNA markers, are discussed with reference to case studies.

10.2 INTRODUCTION

DNA-based assays have revolutionized and modernized our ability to characterize genetic variation. The first advantage of molecular techniques is their capacity to detect genetic diversity at a higher level of resolution than other methods. Furthermore, DNA-based assays are robust and speedy, and information may be obtained from small amounts of plant material at any stage of development. In addition, they are not affected by environmental conditions.

However, managing biodiversity requires more than genetic characterization through DNA polymorphism detection, because information is needed to address key issues of both *ex situ* and *in situ* plant germplasm management and to assist in the process of decision-making. For *ex situ* crop germplasm maintenance, molecular tools may contribute to the sampling, management and development of "core" collections as well as the utilization of genetic diversity. For the *in situ* and on-farm preservation strategies of genetic resources, molecular markers might help in recognizing the most representative populations within the gene pool of a landrace and identifying of the most suitable strategies for their managing and use.

Many different molecular techniques are presently available and each differs in informational content. Multi-locus approaches may be convenient but have some technical and/or analytical drawbacks, such as dominance (i.e. only one allele identified, no possibility to discriminate between homozygous and heterozygous individuals). As a consequence of simultaneous visualization of many marker alleles, multi-locus data typically are analysed as pair-wise comparison of complex patterns that only have meaning relative to others in the same study; thus results are to a limited extent comparable among studies. By contrast, single-locus markers are usually characterized by co-dominance (i.e. both alleles identified in heterozygous individuals) and are therefore more flexible and supply more robust and comparable data (Karp, 2002).

An appropriate use of molecular markers techniques requires clearly defining the issues addressed and the type of information that will be needed (on genetic diversity), and knowing what the different techniques can offer, not only in terms of genetic information, but also resource requirements, reproducibility and adaptability for automation. Furthermore, it is of pivotal importance to consider how the information will be gathered and the way in which the data will be scored and analysed. For accurate and unbiased estimates of genetic diversity, adequate attention has to be devoted to: (i) sampling strategies; (ii) the utilization of various data sets on the basis of understanding their strengths and constraints; (iii) the choice of genetic similarity estimates or distance measures, clustering procedures and other multivariate methods in analyses of data; and (iv) objective determination of genetic relationships (Mohammadi and Prasanna, 2003). For all these reasons, choosing the most appropriate technique may be difficult, hence often a combination of techniques is needed to gather the information of interest.

Up to now most conservation efforts have focused on agriculturally important crops; around one-third of all *ex situ* accessions in gene bank represents just five species: wheat (*Triticum* sp.), barley, rice, maize and beans (*Phaseolus* spp.). The relative over-representation of the five species does not necessarily mean that their genetic diversity has been fully covered (Graner *et al.*, 2004); in fact there is a significant lack of knowledge on the diversity and geographic distribution of less utilized crops as well as their wild relatives (Hammer, Arrowsmith and Gladis, 2003).

This chapter presents two case studies addressed to assess, through molecular tools, the genetic diversity in a minor crop, such as *Cynara cardunculus*, and in landraces of a major crop such as *Zea mays*.

10.3 CASE STUDY 1: ASSESSMENT OF THE GENETIC DIVERSITY IN CYNARA CARDUNCULUS

Cynara cardunculus L. is a diploid ($2n=2x=34$), predominantly cross-pollinated species native to the Mediterranean Basin. It includes two crops: globe artichoke (var. *scolymus* L.) and cultivated cardoon (var. *altilis* DC) as well as wild cardoon (var. *sylvestris* [Lamk] Fiori), a non-domesticated perennial that has been recognized as the ancestor of both cultivated forms. Globe artichoke represents an important component of the South European agricultural economy but is grown all over the world for its large immature inflorescences (capitula); its commercial production is mainly based on perennial cultivation of vegetatively propagated clones. Cultivated cardoon is grown for its fleshy stems and roots and is of some regional importance in Italy, Spain and the south of France; its propagation is carried out through seed.

Previous studies have shown that all the forms within *Cynara cardunculus* have the following characteristics: (i) they are a promising source of seed oil, both with respect to quantity and quality; (ii) they can be exploited for the production of lingo-cellulosic biomass for energy or paper pulp; and (iii) they are a source of biopharmaceuticals. The roots contain inulin, a known improver of human intestinal flora, while the leaves are a source of antioxidant compounds, such as luteolin and di-caffeoylquinic acids (cynarin). Notwithstanding its wide possibilities of exploitation, little is known on the amount and distribution of genetic diversity. In order to assess this, the following questions had to be addressed:

- Which area should be included in the survey?
- How can diversity of natural and cultivated populations growing *in situ* be assessed rapidly and efficiently to quantify the distribution of genetic variation and gather information for both identification of populations representative of the genetic variation and application of sampling or *in situ* preservation strategies?
- How much diversity is present in the *Cynara cardunculus* cultivated forms of different growing regions and what criteria should be adopted for the development of a core collection?
- Can individual plants be adequately identified for future application of plant breeding programmes?

Of course, DNA-based methods can efficiently contribute to answering the above questions, but which are the most suitable for obtaining results that are reproducible by different laboratories, capable of being scored and analysed using standardized methods, and/or suitable for entry into database? Since techniques detecting heterozygotes (i.e. co-dominant markers) and providing data on allelic differences were desirable, microsatellites (simple sequence repeats [SSR]) were considered the most suitable markers although not available at the time of starting the work. Our first objective was thus to develop highly informative microsatellite markers by following different strategies, as reported by Acquadro *et al.* (2003; 2005a; 2005b). Fifty primer pairs were designed and 32 were chosen as being highly polymorphic. We also felt it necessary to complement the microsatellite analysis with amplified fragment length polymorphism markers (AFLPs) due to their high reproducibility and their high information content.

Two spatially isolated areas were chosen for our survey: Sardinia and Sicily, the latter implicated as the origin of artichoke domestication. In both islands, wild cardoon extensively colonizes dry and undisturbed areas while autochthonous globe artichoke germplasm is still the most commonly grown. The first problem to face was how many populations within each area and how many plants within each population had to be sampled. Indeed, in studies aimed at analysis of population structure, it is necessary to balance the need to collect as large a sample size as possible and to get allele frequencies from as many loci as possible, with the need to screen as many populations as possible.

10.3.1 Wild cardoon genetic variation

On the basis of the different pedo-climatic conditions of the growing areas, we identified three populations in Sardinia and four in Sicily (Figure 12) (Portis *et al.*, 2004; 2005a).

It is well known that the minimum sample size for detecting alleles at a given frequency is greater in inbreeding than in outbreeding species (Gregorius, 1980);

FIGURE 12
Geographic location of wild cardoon (*C. cardunculus var. sylvestris*) and globe articoke populations

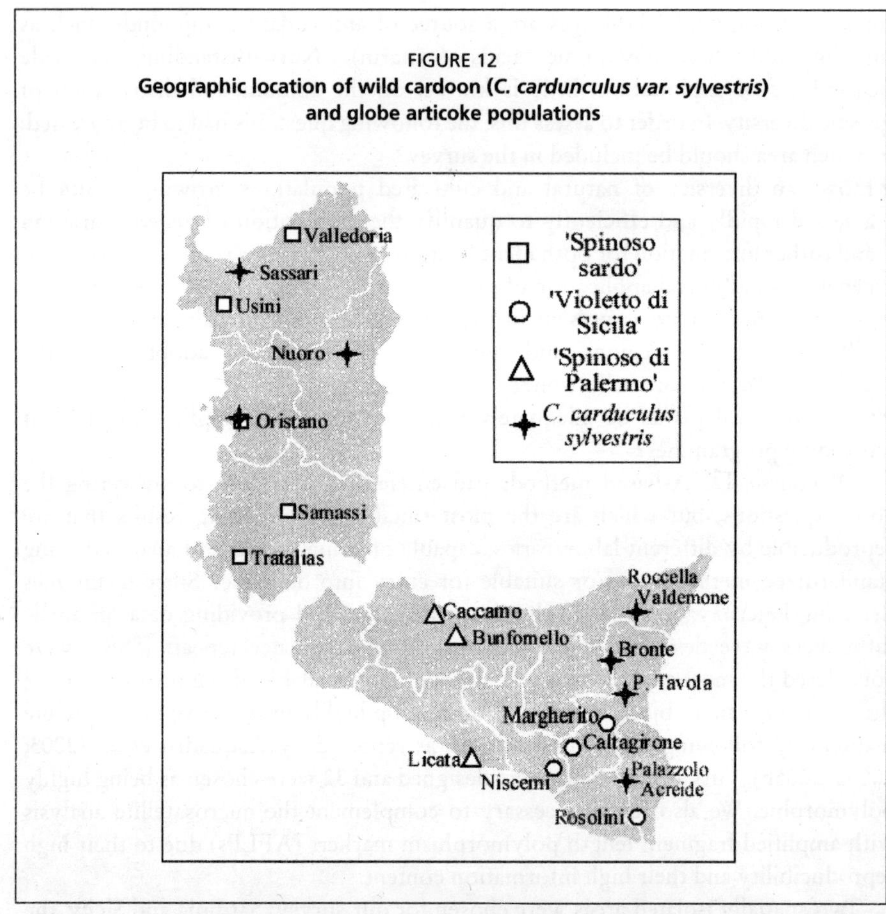

being *Cynara cardunculus*, an allogamous species, 30 individuals per population were randomly sampled, which ensured (P< 0.95) the detection of alleles present at relative frequencies between 0.08 and 0.09 (Gillet, 1999). Plants were genotyped using the developed SSRs and seven AFLP primer combinations, which generated more than 400 polymorphic bands. Genetic divergence between populations was found to be consistent between the two marker systems.

As a result of the geographical isolation, the Sardinian and Sicilian populations were clearly differentiated, forming two distinct gene pools (Figure 13D). Both marker systems show that the Sardinian and Sicilian populations possess a remarkable number of unique (private) alleles (Table 12) showing that the two gene pools must have evolved independently.

Several criteria have been suggested to identify priority population for sampling the maximum possible genetic variation (Maguire, Peakall and Saenger, 2002): (i) allelic richness, or the number of alleles per locus; (ii) evaluated as locally common alleles, that is, those that are common in one to several populations but not in the

FIGURE 13

PCO plot of first three principal coordinates depicting the genetic relationship among accessions from populations of globe artichoke: "Spinoso sardo" (A), "Violetto di Sicilia" (B), "Spinoso di Palermo" (C), and wild cardoon (D)

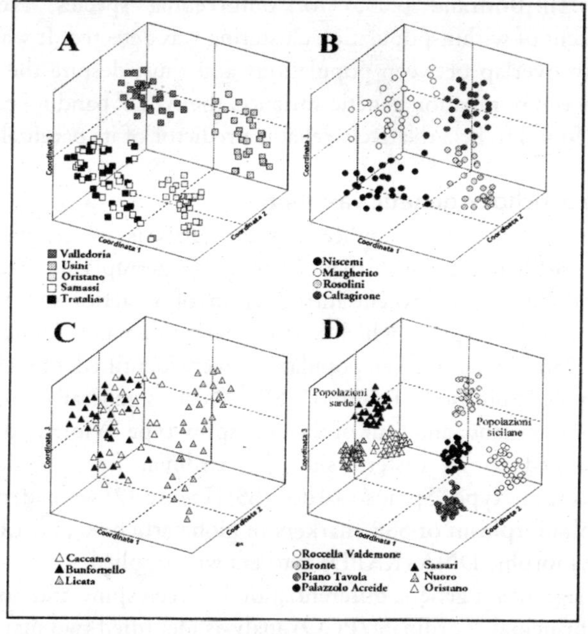

TABLE 12

Number of bands, and common and rare alleles within a location, detected with SSR and AFLP. Roccella (ROC), Bronte (BRO), Piano Tavola (TAV), Palazzolo (PAL), Sassari (SAS), Nuoro (NUO), Oristano (ORI)

Population	No. polymorphic bands		No. locally common alleles		No. private alleles	
	SSR	AFLP	SSR	AFLP	SSR	AFLP
BRO	35	342	7	22	3	6
PAL	40	347	10	25	5	5
ROC	34	302	6	15	3	5
TAV	24	328	4	13	2	3
ORI	34	286	6	16	1	5
SAS	25	269	3	12	1	4
NUO	22	250	2	13	0	4
Sicilian populations	60	419	26	70	2	22
Sardinian populations	42	333	10	36	3	19

species as a whole; and (iii) identified as unique (private) alleles. Based on any one of these criteria and for both AFLP or SSR data sets (Table 12), the same priority populations were identified: Palazzolo and Bronte in Sicily, and Oristano in Sardinia.

In both islands, most of the AFLP and SSR genetic variation was present within rather than between populations, which is consistent with data reported by Gaudel, Taberlet and Till-Bottraud (2000) for outbreeding species. Nevertheless, a remarkable extent of within-population clustering was detected. It was not possible to identify any overlap between populations and thus, despite the high ratio of within-to-between population genetic variance, the AFLP banding pattern of each genotype was found to be a relatively reliable predictor of its parental population.

10.3.2 Globe artichoke genetic variation

An analogous approach was applied for the analysis of genetic variation in autochthonous Sicilian and Sardinian globe artichoke germplasm, which is at risk of genetic erosion due to the recent introduction of varieties selected abroad or germplasm from other regions which best fit market demand (Lanteri *et al.*, 2001; Portis *et al.*, 2005b). In Sicily seven populations were identified, of which three were of the spiny type, "Spinoso di Palermo" (SP), which is generally cultivated on the western side of the island, and four of the non-spiny type "Violetto di Sicilia" (VS), which is confined to its eastern side. In Sardinia, five populations of the autochthonous spiny type "Spinoso sardo" (SS) (Figure 12) were identified. Due to the limited polymorphism of SSR markers in globe artichoke, AFLP and random amplified polymorphic DNA (RAPD) markers were applied.

In Sicily a significant genetic differentiation between spiny and non-spiny types was found as principal coordinate (PCO) analysis identified two main clusters, one of which groups the three representative SP populations, while the other groups the four VS populations (data not reported). Around one-third of the AFLPs scored were found specific to one or other of these two types.

In both Sicilian and Sardinian populations, most the genetic variation was present within, rather than between populations. As the three varietal types are vegetatively propagated, the within-genetic variation presumably reflects their multiclonal composition, a direct consequence of the limited selection criteria adopted by farmers. An additional source of variation may be via spontaneous mutations, which are maintained as they are not subject to any meiotic sieve and might not be detectable.

Despite the high level of within-population genetic variation present, most of the populations could be genetically differentiated from one another due to farm fragmentation or adaptation to local pedo-climatic conditions. In Sicily the between-population differentiation was more evident in VS, allowing the PCO to define four clusters with minimal overlap (Figure 13B). The opposite was true in Sicilian SP and in Sardinian SS: two of the populations were partially overlapping, presumably as a consequence of some exchange of material between farmers (Figure 13A-C). Our data resulted very informative for the implementation of on-farm germplasm preservation strategies. On the basis of the criteria reported above, the Rosolini (VS) and the Buonfornello (SP) populations in Sicily and the Oristano population in Sardinia are the most representative of the gene pool of the varietal types, and priority should be given to them for application of on-farm preservation strategies.

10.3.3 Globe artichoke and cultivated cardoon germplasm characterization

The use of molecular markers that quantify the genetic diversity within and between accession may significantly increase the efficiency of the assessment and management of germplasm collection by reducing redundancy. Several works have detailed the philosophy behind optimizing collections to ensure diverse genetic representation through either the creation of "core" collections or some form of hierarchical sampling. We applied eight AFLP primer combinations for characterizing a living collection of globe artichoke germplasm maintained at CRAS (Oristano, Sardinia).

We also included in the analysis accessions collected in the field in Spain, Turkey and the United States for a total of 89 varietal types; for three of them different provenances were assessed for a total of 118 accessions (Lanteri *et al.*, 2004a). Figure 14 shows the dendrogram from unweighted pair-group method arithmetic average (UPGMA) cluster analysis of AFLP data. Our results suggest that traits selected by humans play an important role in understanding variation and differentiation within cultivated artichoke germplasm. Two main clusters were found within branch A: A1 and A2. Cluster A1 mainly contained Catanesi types with small elongated heads, while cluster A2 included the Romaneschi types, with big spherical or sub-spherical non-spiny heads, mainly cultivated in central Italy together with the American accessions of Green Globe, which accounts for more than 85 percent of the artichoke production in the United States. Branch B contained two additional clusters: B1 and B2, both of which included Spinosi and Violetti types with medium-small heads. Cluster B1 also included all the Turkish accessions. AFLP data furnished important information for assembling a core collection of globe artichoke germplasm, taking into account the hierarchical structure of the gene pool. Our results showed that the genetic variation detected within the same varietal type was in some cases higher than that found between varietal types. The Jaccard's similarity index among clones of the same varietal type might thus be considered a threshold value that identifies material sharing the same genetic background. On the basis of this threshold a limited number of core subset could be identified. Genetic studies in selected crops have demonstrated that widespread and localized alleles occurring in the entire collection are usually contained in the core subset, with only rare localized alleles excluded (van Hintum *et al.*, 2000). The core subset can thus provide an introduction to further studies on biodiversity of the entire collection or to identify suitable material for future breeding efforts. Furthermore, the results obtained in this study, as well as in another study aimed at genotyping selected clones of the varietal type "Spinoso sardo" (Lanteri *et al.*, 2004b), showed that AFLP markers are useful to identify duplicates and provide evidence for the uniqueness of a particular genotype.

FIGURE 14

Dendrogram obtained from UPGMA cluster analysis of AFLP data of 118 globe artichoke accessions

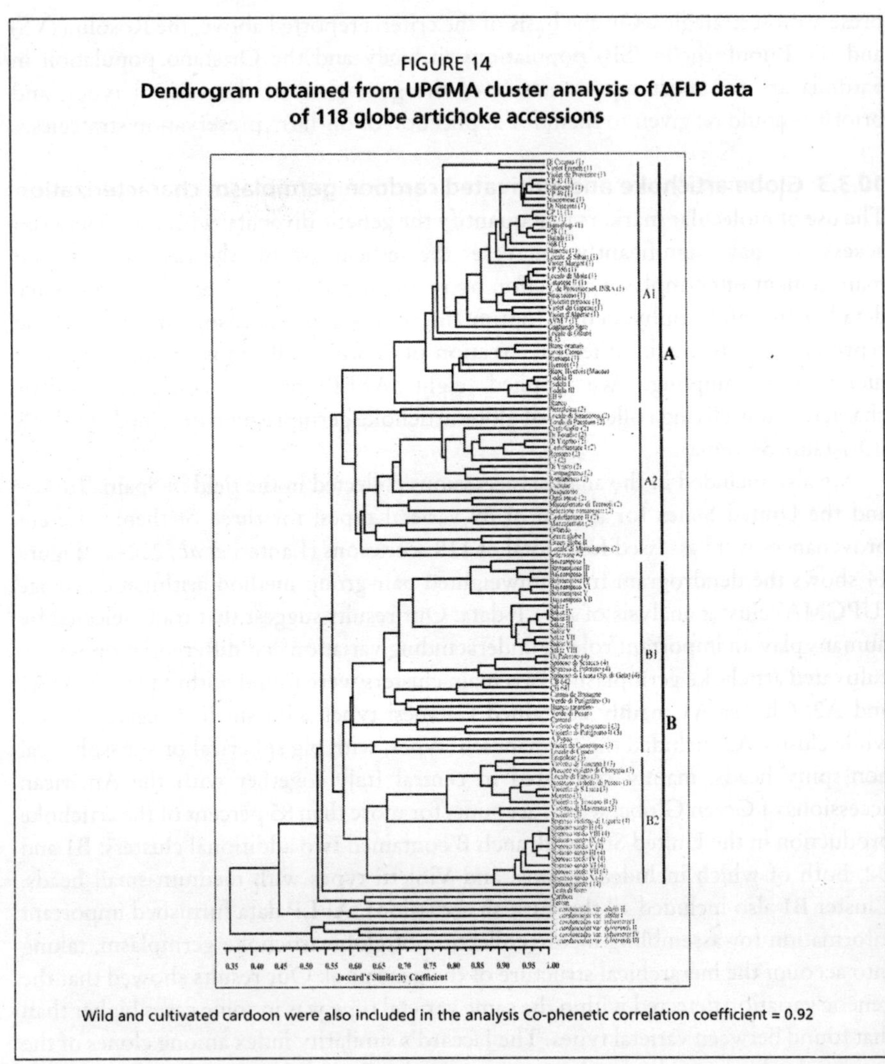

Wild and cultivated cardoon were also included in the analysis Co-phenetic correlation coefficient = 0.92

An analogous study was carried out in cultivated cardoon (Portis *et al.*, 2005c), whose genetic variation of the material in cultivation in Spain and Italy was assessed by DNA profiling at five microsatellite loci and with eight AFLP primer combinations. The analysis of genetic similarities showed that the Spanish and Italian accessions represent two distinct gene pools. This study also demonstrated that a fruitful method in characterizing germplasm collections with AFLP markers is to use a two-tiered approach: first, using low-density profiles to compare all samples and resolve the main cluster, and then using high-density profiles to resolve samples within clusters.

10.4 CASE STUDY 2: ITALIAN LANDRACES OF MAIZE AND MOLECULAR MARKERS FOR THEIR CHARACTERIZATION AND CONSERVATION

Maize (*Zea mays* L.) is one of the most important crops in Italian agriculture. The species was introduced in the national cultivation system approximately four centuries ago and is grown mainly for human consumption. Since then a number of landraces have been developed to meet specific needs of cultivation and utilization and to overcome environmental constraints of different areas. As a consequence, new landraces originated from the original populations introduced, through adaptation to local conditions as well as hybridization brought about by continuous exchange and trade. These landraces were locally maintained by farmers as open-pollinated populations, thus each represented a collection of highly heterozygous and heterogeneous plants. Although a considerable range of variation within each population was present, a between-population differentiation was detectable for several distinctive traits as a consequence of both natural and human selection pressure.

Within the last few decades, the Italian agricultural scenery has profoundly changed and the subsistence mixed farming unit is now transformed into an intensive monoculture (Bertolini *et al.*, 1998). At present, a small number of populations of flint maize (*Z. mays* var. *indurata*) can be found under very peculiar agricultural situations or in marginal areas, such as alpine valleys, and on small fields traditionally managed according to low-input agronomic practices, and with production exclusively addressed to human consumption (Lucchin, Barcaccia and Parrini, 2003). The agricultural environment together with the traditional diet of these regions allows preservation of some landraces and limits diffusion of modern hybrids. Unfortunately, many locally cultivated populations were lost before it was realized that they were important sources of germplasm. Maize breeders have recently become more aware of the need for both maintaining genetic diversity among hybrid varieties and improving the management of genetic resources through the conservation of landraces. Consequently, there was renewed interest for *in situ* conservation of the landraces, not only to preserve important sources of genetic material for breeding, but also to allow their valorization as essential components of sustainable agriculture. Landraces are the cultivated maize material with the highest genetic variation as well as with the best adaptation to the natural and anthropological environment where they have evolved. In addition, they contain locally adapted alleles and likely represent an irreplaceable bank of highly co-adapted genotypes. Knowledge of genetic diversity among local populations and breeding stocks is expected to have a significant impact on the improvement of this crop. In maize, information on both qualitative and quantitative morphological traits of existing landraces may be useful in maintaining their genetic variability and preserving them from genetic erosion. Nowadays, after years of lack of interest in the so-called "old local varieties", this valuable source of maize germplasm has been rediscovered and exploited as a niche crop suitable for the cultivation of marginal lands.

The development of molecular markers has greatly facilitated basic and applied research programmes of maize genetics and breeding. DNA polymorphism assays are also known as powerful tools for characterizing gene pools and investigating germplasm resources.

10.4.1 Molecular characterization of field populations belonging to an Italian landrace of flint maize (*Zea mays* var. *indurata*)

A comparative characterization of farmer populations of the flint maize landrace "Nostrano di Storo" was recently carried out by Barcaccia, Lucchin and Parrini (2003) using different types of PCR-based markers. The inbred line B37 and three synthetics (VA143, VA154 and VA157) selected from as many landraces were used as reference standards. Genetic diversity and relatedness were evaluated with SSR and Inter-SSR markers. Nei's total genetic diversity as assessed with SSR markers was HT=0.851 while the average diversity within populations was HS=0.795. The overall Wright's fixation index FST was as low as 0.066. Thus, more than 93 percent of the total variation was found within population. Dice's genetic similarity coefficients within and between populations on the basis of Inter-SSR fingerprints were 0.591 and 0.564, respectively. The UPGMA dendrogram displayed all populations except for one clustered into a distinct group, in which a synthetic variety selected from the landrace "Marano Vicentino" was also included (Figure 15).

One population and the other two synthetics, "Spino Bresciano" and "Dente di Cane Piemontese", were clustered separately. Findings suggest that although a high variability can be found among plants, most of their genotypes belong to the same landrace locally called "Nostrano di Storo" (Barcaccia, Lucchin and Parrini, 2003). This result was also confirmed by a further molecular investigation carried out using AFLP and RAPD markers to fingerprint pooled DNA samples from all farmer populations.

FIGURE 15
UPGMA dendrogram of the maize farmer populations

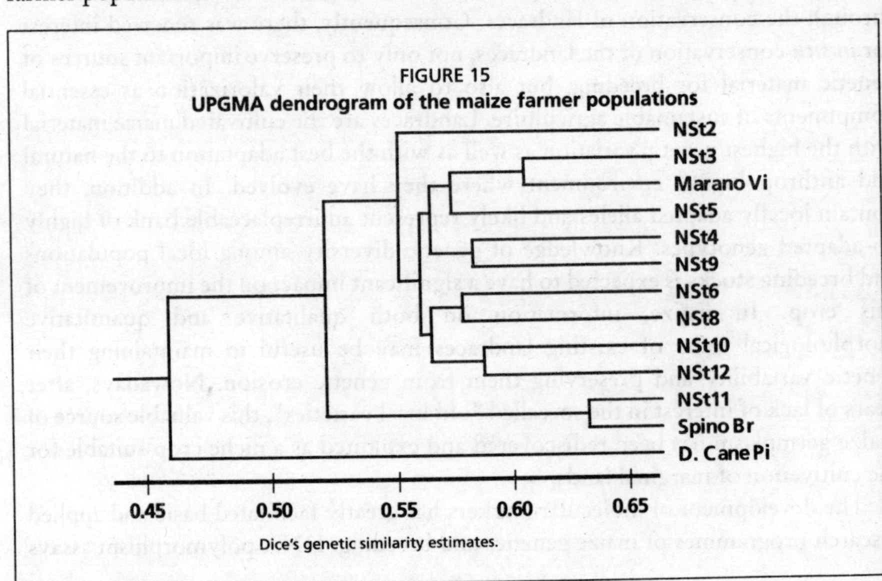

Although gene flow from commercial hybrids might have occurred, the large number of polymorphisms and the presence of both unique alleles and alleles unshared with B37 and synthetics are the main factors underlying the value of this flint maize landrace as a source of genetic variation and peculiar germplasm traits. Because of its exclusive utilization for human consumption, such a molecular marker characterization will be a key step to promoting the *in situ* conservation and protection of the landrace.

10.4.2 Construction of a linkage map for a maize landrace based on a pseudo-testcross strategy using multi-locus PCR-based markers

Genetic linkage maps based on molecular markers represent basic tools for investigating and characterizing local germplasm resources. A linkage map of the flint maize landrace "Nostrano di Storo" based on dominant multi-locus PCR-derived markers (RAPD, I-SSR, AFLP and SAMPL), was constructed according to a one-way pseudo-testcross mapping strategy (Barcaccia *et al.*, 2000; 2005). The feasibility of such a study depended on the presence of high levels of heterozygosity in the landrace parent mapped and on the informativeness of the PCR-based marker systems used. Co-dominant single-locus SSR markers were adopted to assign each linkage group to a specific chromosome and all marker loci to specific chromosome arms. The final genetic map includes 282 marker loci and covers 1.826 cM (Table 13).

This genetic map based on multi-locus marker systems and on easily detectable molecular markers will find application and prove useful for rapidly characterizing the genetic diversity within and relatedness among farmer populations belonging to the "Nostrano di Storo" landrace maintained according to different conservation strategies.

TABLE 13
Summary of the map statistics, including length, number of marker loci per chromosome and total

	Chromosome										
Statistics	**1**	**2**	**3**	**4**	**5**	**6**	**7**	**8**	**9**	**10**	**Total**
Map length (cM)	217	220	166	260	172	161	173	150	142	165	1,826
No. of AFLPs	29	27	27	20	24	19	25	14	20	17	222
No. of SAMPLs	1	6	6	1	0	2	2	2	0	4	24
No. of RAPDs	0	0	3	1	0	2	0	1	0	0	7
No. of Inter-SSRs	1	2	0	1	0	0	0	0	1	0	5
No. of SSRs	3	2	2	2	4	3	2	2	2	2	24
Total marker loci	34	37	38	25	28	26	29	19	23	23	282
Average map density	6.4	6.0	4.4	10.4	6.2	6.2	6.0	7.9	6.2	7.2	6.5

10.4.3 Assessment of the optimal plant and molecular marker sample size to estimate genetic diversity in maize landraces

The genetic characterization of landraces represents an essential step for their conservation. This requires establishing the most appropriate system and type of markers (i.e. random or mapped) and the minimum number of markers and plants required to describe the genetic structure of a given population. In spite of their importance for the success of any germplasm conservation programme, little information is available on landraces because almost all studies were performed on inbred lines. A sample of plants were chosen as representative of the landrace in terms of morpho-phenological and agronomic traits, and then assayed at hundreds of marker loci, either mapped or random. Genetic similarity and diversity coefficients computed using mapped markers proved to be significantly higher than estimates based on total markers. Moreover, no significant changes of marker allele frequency and polymorphism information content were scored when the number of sampled ears was progressively reduced from 50 to 15, even if a steady increase of standard errors was observed. The influence of the number and type of molecular markers on genetic similarity and diversity measurements was also investigated: no significant changes in terms of mean polymorphic index content (PIC) values were observed, whereas the coefficient of variation (CV) of standard deviations raised proportionally to the reduction of sample size. A total of 120 random markers and 80 mapped markers were needed to get CVs of standard deviations lower than 5 percent. Also, when the number of ears and markers were evaluated together, no significant changes of the mean PIC values were observed and an increase of the molecular data variability was confirmed (Pallottini, 2002).

Influence of marker and plant sampling on statistics used to measure genetic diversity should be useful to investigate the genetic consequences of different modes of conservation of maize landraces (on-farm, *in situ* and *ex situ*). In conclusion, reliable and effective investigations of landrace population genetic structure with AFLP markers can be performed using at least 30 ears per population and one plant per ear, and require at least 80 mapped marker loci.

10.4.4 Effects of different conservation strategies (on-farm, *in situ* and *ex situ*) on the population genetic structure of maize landraces as assessed with molecular markers

The on-farm, *in situ* and *ex situ* conservation methods may exert a different influence on the genetic structure of populations grown by farmers. This influence should be accurately evaluated to avoid genetic erosion and conservation programme failure. In fact, the loss of genetic diversity could be due to inbreeding that can result from drift and migration to natural and human selection and gene flow. Each of these factors has a different relative importance on the types of conservation methods. Molecular markers were used to investigate the influence of the conservation strategy on the genetic structure of farmer populations grown for two years with three different methods: (i) on-farm conservation by farmers, using own seeds and traditional agronomic practices; (ii) *in situ* conservation in the original area but taking into account the spatial isolation from other fields cultivated with hybrid varieties; and (iii) *ex situ* conservation far from the original area with no gene flow due to the total absence of fields grown with the same crop (Pallottini, 2002). Statistical tests failed to reveal any significant difference in terms of diversity/similarity absolute values among the populations conserved according to the three distinct strategies (Figure 16).

However, about 10 percent of the comparisons performed for the marker allele frequency parameter at the total assayed loci showed significant differences. Even the differences between genetic variation parameters computed for mapped and random marker loci were significant. In particular, some marker loci were more affected than others by changes of the marker allele frequency depending on the conservation method. These markers, distributed throughout the genome, could be related to important genes involved in the adaptation to environmental conditions or responsible for traits evaluated in the selection by farmers.

In sum, although all conservation methods studied have determined the significant changes to the genetic structure of the farmer populations, the genetic variation and diversification that occurred with *ex situ* conservation was much stronger than that observed for *in situ* and on-farm conservation. It is worth mentioning that to monitor these changes, the level at which the investigation is performed is essential. When the mean values of the more common genetic diversity and/or similarity indexes are taken into account, no significant differences are highlighted because of the large set of molecular data and the occurrence of bidirectional changes of marker allele frequencies over all marker loci. Consequently, variation of the marker allele frequency has to be computed and interpreted at each single marker locus or between pairs of marker loci, but not on the whole molecular marker data set.

FIGURE 16

PIC values and standard errors computed in the original population (or) and in the populations obtained from on-farm (of), *in situ* (is) and *ex situ* (es) conservation strategies, using mapped and random markers

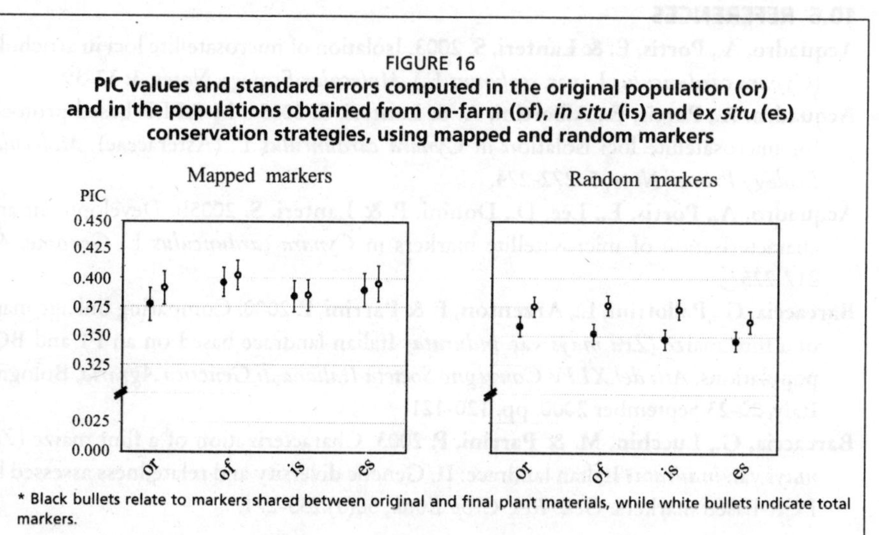

* Black bullets relate to markers shared between original and final plant materials, while white bullets indicate total markers.

10.5 CONCLUSIONS

These case studies show how molecular marker techniques may help in characterizing and managing genetic diversity. However, other questions have to be answered. Molecular tools are very informative but are generally employed anonymously. They are able to detect high levels of DNA polymorphisms, but are they really providing the kind of information required to make effective and sound judgments on diversity? What is the functional relevance of the polymorphism detected? Indeed, understanding the significance or assessing the value of the diversity is still a difficult challenge.

New approaches have been recently developed for adapting the current PCR-based techniques to target functional diversity. As stated by Tanksley and McCouch (1997), "new findings from genome research indicate that there is a tremendous genetic potential locked up in germplasm collections that can be released only by shifting the paradigm from searching for phenotypes to searching for superior genes with the aid of molecular linkage map." At present, the increasing information available from genome mapping means that markers known to be very closely linked to traits of interest can be better addressed for characterizing genetic diversity and help in identifying variation of use to breeders. Furthermore, the identification of genes controlling a trait and the availability of their DNA sequences may facilitate the classification of variation in germplasm pools. High-resolution genetic maps enable closely linked markers to be used and the increasing numbers of expressed sequence tags (ESTs) and single nucleotide polymorphism (SNP) markers provide routes for more targeted sequence-based approaches. Classification of the sequence variants at a targeted locus would substantially reduce the amount of work needed to assess their potential for breeding and lead to the identification of superior alleles (Sorrels and Wilson, 1997).

10.6 REFERENCES

Acquadro, A., Portis, E. & Lanteri, S. 2003. Isolation of microsatellite loci in artichoke (*Cynara cardunculus* L. var. *scolymus* L.). *Molecular Ecology Notes*, 3: 37-39.

Acquadro, A., Portis, E., Albertini, A. & Lanteri, S. 2005a. M-AFLP-based protocol for microsatellite loci isolation in *Cynara cardunculus* L. (Asteraceae). *Molecular Ecology Primer Note*, 5: 272-274.

Acquadro, A., Portis, E., Lee, D., Donini, P. & Lanteri, S. 2005b. Development and characterisation of microsatellite markers in *Cynara cardunculus* L. *Genome*, 48: 217-225.

Barcaccia, G., Pallottini, L., Arzenton, F. & Parrini, P. 2000. Comparing linkage maps of a flint maize *(Zea mays* var. *indurata)* Italian landrace based on an F1 and BC1 populations. *Atti del XLIV Convegno Società Italiana di Genetica Agraria,* Bologna, Italy, 20-23 September 2000. pp. 120-121.

Barcaccia, G., Lucchin, M. & Parrini, P. 2003. Characterization of a flint maize (*Zea mays* var. *indurata*) Italian landrace: II. Genetic diversity and relatedness assessed by PCR-based markers. *Gen. Res. Crop Evol.*, 50(3):253-271.

Barcaccia G., Pallottini, L., Parrini, P. & Lucchin, M. 2005. A genetic linkage map of a flint maize (*Zea mays* var. *indurata* L.) Italian landrace using a one-way pseudo-testcross strategy and multi-locus PCR-based markers. *Maydica.* (in press)

Bertolini, M., Bianchi, A., Lupotto, E., Salamini, F., Verderio, A. & Motto, M. 1998. Maize. *In* G.T. Scarascia Mugnozza & M.A. Pagnotta, eds. *Italian contribution to plant genetics and breeding.* Viterbo, Italy. Quatrini A. & F. Publisher. pp. 209-229.

Gaudel, M, Taberlet, P. & Till-Bottraud, I. 2000. Genetic diversity in an endangered alpine plant, *Eryngium alpinum* L. *(Apiaceae)*, inferred from amplified fragment length polymorphism markers. *Mol. Ecology,* 9: 1625-1637.

Gillet, E.M. 1999. *Sampling strategies for marker analysis.* (available at www.webdoc.sub.gwdg.de/ebook/y/1999/whichmarker.htm).

Graner, A., Dehmer, K.J., Thiel, T. & Borner, A. 2004. Plant genetic resources: benefits and implications of using molecular markers. *In* M.C. de Vicente, ed. *The evolving role of genebanks in the fast-developing field of molecular genetics.* Rome, IPGRI. pp. 26-32.

Gregorious, H.R. 1980. The probability of losing an allele when diploid genotypes are sampled. *Biometrics*, 36: 632-652.

Hammer, K., Arrowsmith, N. & Gladis, T. 2003. Agrobiodiversity with emphasis on plant genetic resources. *Naturwissenschaften*, 90: 241-250.

Karp, A. 2002. The new genetic era: Will it help us in managing genetic diversity? *In* J.M.M. Engels, R.V. Ramanatha, A.H.D. Brown & M.T. Jackson, eds. *Managing Plant Genetic Diversity.* Rome, IPGRI. pp. 43-56.

Lanteri, S., Di Leo, I., Ledda, L., Mameli, M.G. & Portis, E. 2001. RAPD, variation within and among populations of globe artichoke (*Cynara scolymus* L.), cv "Spinoso sardo". *Plant Breed,* 120 (3): 243-246.

Lanteri, S., Saba, E., Cadinu, M., Mallica, G.M., Baghino, L. & Portis, E. 2004a. Amplified fragment length polymorphism for genetic diversity assessment in globe artichoke. *Theor. Appl. Genet.,* 108: 1534-1544.

Lanteri, S., Acquadro, A., Saba, E. & Portis, E. 2004b. Molecular fingerprinting and evaluation of genetic distances among selected clones of globe artichoke (*Cynara scolymus* var. *cardunculus* L.) 'Spinoso sardo'. *J. Hortic. Sci. Biotech.,* 79(6): 863-870.

Lucchin, M., Barcaccia, G. & Parrini, P. 2003. Characterization of a flint maize (*Zea mays* var. *indurata*) Italian landrace: I. Morpho-phenological and agronomic traits. *Gen. Res. Crop Evol.,* 50(3):315-327.

Maguire, T.L., Peakall, R. & Saenger, P. 2002. Comparative analysis of genetic diversity in the mangrove species *Avicennia marina* (Forsk.) Vierh. (*Avicenniaceae*) detected by AFLPs and SSRs. *Theor. Appl. Genet.,* 104: 388-398.

Mohammadi, S.A. & Prasanna, B.M. 2003 Analysis of genetic diversity in crop plants - salient statistical tools and considerations. *Crop. Sci.,* 43: 1235-1248.

Pallottini, L. 2002. Variation of the genetic structure of populations in relation to different strategies of germplasm conservation (on farm, *in situ* and *ex situ*) in local varieties of maize: the case study of the "Nostrano di Storo" (*Zea mays* var. *indurate* L.). Padua, Italy, University of Padua. pp. 316. (PhD thesis)

Portis, E., Mauromicalem, G., Cominom, C., Acquadrom, A. & Lanterim, S. 2004. Amplified fragment length polymorphism (AFLP) analysis of genetic variation in Sicilian populations of wild cardoon (*Cynara cardunculus* L. var. *sylvestris*). *Acta Horticulturae*, 660: 229- 234.

Portis, E., Acquadro, A., Comino, C., Mauromicale, G., Saba, E. & Lanteri, S. 2005a. Genetic structure of island populations of wild cardoon (*Cynara cardunculus* L. var. *sylvestris* [Lamk] Fiori) detected by AFLPs and SSRs. *Plant Science*, 169: 199-210.

Portis, E., Mauromicale, G., Barchi, L., Mauro, R. & Lanteri, S. 2005b. Population structure and genetic variation in autochthonous globe artichoke germplasm from Sicily Island. *Plant Science*, 168: 1591-1598.

Portis, E., Barchi, L., Acquadro, A., Macua, J.I. & Lanteri, S. 2005c. Genetic diversity assessment in cultivated cardoon by AFLP (amplified fragment length polymorphism) and microsatellite markers. *Plant Breed*, 124: 299-304.

Sorrels, M. & Wilson, W.A. 1997. Direct classification and selection of superior alleles for crop improvement. *Crop Sci.*, 37: 691-697.

Tanksley, S.D. & McCouch, S. 1997. Seed banks and molecular maps: unlocking genetic potential from the wild. *Science*, 277: 1063-1066.

van Hintum, Th.J.L., Brown, A.H.D., Spillane, C. & Hodgkin, T. 2000. Core collections of plant genetic resources. *IPGRI Technical Bulletin* 3. Rome, IPGRI.

11. Molecular analysis of gene banks for sustainable conservation and increased use of crop genetic resources

Marcio Elias Ferreira

11.1 SUMMARY

The effective use of crop genetic resources stored in gene banks by breeding programmes is limited. The number of accessions deposited in gene banks, however, is continuously growing. Slow germplasm characterization has been pointed out as a major cause of this discrepancy. Molecular marker technology offers opportunities to increase the use of crop genetic resources deposited in gene banks. High throughput genotyping of germplasm accessions allows for the examination of genetic relationships and sampling of core collections representative of the allelic richness of the gene bank. Core collections can be used for intensive phenotypic evaluation of traits of agronomic importance and re-sequencing of candidate genes associated with their control. Single nucleotide polymorphism (SNP) variation between the accessions of the collection can be associated with phenotypic variation. The integration of genomic technology and the characterization of germplasm banks will play an important role in the sustainable conservation and increased use of crop genetic resources.

11.2 INTRODUCTION

The conservation, management and use of germplasm maintained in gene banks poses a number of challenges to the researchers dedicated to the investigation of plant genetic resources. Common problems include, for example, the development of strategies for sampling representative individuals in natural populations, the improvement of tools and technology for long-term conservation or high throughput genetic analysis of an increasingly high number of stored acessions. Central to sustainable conservation is the knowledge of the genetic diversity present in a gene bank. This is also key to the potential exploitation of gene banks by breeding programmes. Therefore, the characterization of the accessions maintained in the collection and the examination of the genetic relationship between them is important for the sustainable conservation and increased use of crop genetic resources. Germplasm characterization of plant accessions deposited in gene banks has been limited and is probably a major cause for the limited use of accessions in breeding programmes.

Germplasm characterization refers to the observation, measurement and documentation of heritable plant traits in a collection. The resulting data allows for identifying and classifying accessions, and building a catalogue of descriptors with embedded biological information that is essential for collection management or for direct use in agriculture. The characterization of plant germplasm, therefore, aims at describing and understanding the genetic diversity of the organisms under study. Today, germplasm characterization has been developed based mostly on morphological descriptors, agronomic descriptors (traits) and molecular marker technology.

The current characterization of plant germplasm collections relies strongly on morphological descriptors. Morphological descriptors are reliable, easy to study and relatively low cost to evaluate. However, the use of morphological descriptors presents some limitations:

(i) limited polymorphism, lowering the potential success of an extended classification approach, which would require a high number of descriptors in order to compensate for the small number of morphotypes;

(ii) potential environmental influence on the phenotype, making the process of evaluation and information exchange even more complex. Here one should be careful with false positives when the environment affects specific morphotypes;

(iii) the impact of a morphological descriptor in the viability of the individual.

Germplasm characterization based on agronomic traits, on the other hand, is particularly useful in crops of economic importance. The amount of data related to agronomic traits that is available by crop germplasm evaluation is limited. Due to reasons that include relatively high costs and the difficulties of large-scale experimental trials, the use of agronomic evaluation to characterize germplasm collections is far from the actual need of uncovering the phenotypes of agronomic interest in accessions of a collection. More should definitely be done in this area to stimulate a higher use of stored germplasm in breeding programmes. A complete agronomic trait evaluation of crop germplasm in the next few years, seems to be pratically impossible to achieve, however.

Recently, germplasm characterization based on molecular methods has experienced great development. The methods that reveal sequence polymorphism in the DNA structure, known as molecular markers, fomented a revolution in the speed and quality of large-scale plant germplasm characterization. The routines of germplasm characterization have relied increasingly on the use of such methods. Among all different classes of molecular markers available for evaluating genetic diversity, microsatellites or simple sequence repeats (SSRs) (Tautz, 1989; Weber and May, 1989) are well known for their potentially high information content and versatility as molecular tools (Ferreira and Grattapaglia, 1996). Hundreds of microsatellite markers have been developed for different species, having their chromosomal location and polymorphism levels determined. The use of fluorescently labelled microsatellite marker panels greatly increased the capacity of semi-automated genotyping of a large number of accessions, allowing for a faster and highly informative characterization of accessions deposited in gene banks.

The objective of this chapter is to discuss how molecular analysis of gene banks can impact on sustainable conservation and increased use of crop genetic resources.

11.3 THE BRAZILIAN UPLAND RICE GENE BANK: A CASE STUDY

Classical attempts to directly use accessions deposited in germplasm banks in breeding programmes have been somehow limited to the identification of sources of genes of interest, such as resistance to plant pathogens or pests, and the transfer of those genes to elite material of the programme. Linkage drag has usually restrained the breeder from the initiative of using accessions from the germplasm bank since the advanced material of the breeding programme is far more attractive than the inherent risk of using germplasm of unknown pedigree, performance or phenotypic adaptation. When the risk is taken, the accessions are usually submitted to a screening condition that reveals the presence of a gene of interest and typically a backcross programme is initiated to tranfer the gene to an elite line or cultivar. This procedure, however, is usually limited to traits of monogenic control. Complex traits require more elaborate methods, such as the inbred backcross line (Wehrhahn and Allard, 1965) or the advanced backcross quantitative trait locus (AB-QTL) mapping, which integrates the classical backcross method with linkage information based on molecular markers (Tanksley and Nelson, 1996). This procedure has been effective to demonstrate how to transfer quantitative trait loci (QTL) identified in wild relatives deposited in gene banks to elite lines or cultivars of a cultivated species (Xiao *et al.*, 1996; Brondani, Rangel and Ferreira, 2002).

More recently, genome sequencing opened the possibility of finding candidate genes for complex traits (candidate QTLs) in the genome of a crop species, and using the germplasm bank, identifying genes of agronomic importance. Marker technology allied with detailed phenotypic characterization of germplasm banks can therefore be potentially useful in gene discovery. The use of this approach is based on the integration of knowledge of genetic diversity of the germplasm bank, genomic information available for the chosen crop and intensive phenotyping of the trait of choice. A case study on how molecular analysis of gene banks can have an impact on sustainable conservation and increased use of crop genetic resources is discussed below using rice as a model.

Rice has one of the largest *ex situ* germplasm collections in the world, comprised of accessions of cultivated (*Oryza sativa* L. and O. *glaberrima* Steud.) and wild species. Because of its widespread use in the planet, *O. sativa* is considered the most important cultivated species of rice. Rice is a daily staple food in the diet of billions of people in the world, including millions of Brazilians. The production and consumption of rice in Brazil is comparable to figures observed in some Asian countries. About 60 percent of the rice production in the country is irrigated (*O. sativa* spp. *indica*), while upland rice (*O. sativa* spp. *japonica*) accounts for 40 percent of the total. A national gene bank of *japonica* rice is maintained by Embrapa. This gene bank contains approximately 4 000 accessions, including landraces collected in villages and isolated rural areas of the country, where rice has been cultivated since its introduction in Brazil centuries ago. This germplasm is a source of genes that control traits of economic importance, such as drought tolerance and resistance to plant pathogens. Most of these rice accessions have not yet been studied in depth. The phenotypic and genotypic characterization of the collection is just beginning.

The traits of economic interest in plants almost always involve a complex genetic control, generally determined by various genes, and a strong interaction with the environment. The genetic control of drought tolerance in plants, for instance, which is a trait of great importance for the genetic improvement of rice and other grass species, is typically quantitative. Several QTLs showing different effects on the phenotypic variation are involved in the control of drought tolerance in rice. Some QTLs might have a strong or major effect on the phenotype; others might have a small but still significant effect on the trait. The genetic and physical mapping of the genome, based on the use of molecular markers and cloned segments of bacterial artificial chromosomes (BACs), has allowed for great advances in the understanding of the genetic control of quantitative traits, at times making it possible to isolate the genes associated with quantitative trait control by the positional cloning approach (Tanksley, Ganal and Martin, 1995). This approach overcame the classic problem of lack of information regarding the gene product of loci of economic interest (usually a quantitative trait), which is very useful when reverse genetics is possible. In other words, lack of knowledge of the protein related to a trait could be circumvented in gene isolation studies using detailed genetic mapping of the region around the locus of interest, followed by sequencing selected clones of large genomic insert libraries and the identification of the desired gene by complementation studies. Genetic mapping allows for the breakdown and examination of the genetic control of a complex trait, making it possible to identify the number of genes that are involved in the expression of a certain trait, to locate these genes or gene regions on the chromosomes, measure their impact on the phenotypic variation, and understand gene interaction in the manifestation of the trait.

An alternative to positional cloning for the isolation of genes controlling quantitative traits has been advocated with the availability of massive DNA sequence data obtained for some species by whole genome sequencing projects. The approach relies on the power of association tests to verify the correlation of sequence variation at candidate genes with phenotypic variation for the trait of interest (Risch, 2000). Recent developments in the area of human genetics, particularly in the identification of factors that cause genetic diseases, have been based on genetic association tests. These tests have been used for many years in genetic studies, especially in classical medical genetics, but the genomic information gathered recently affords an opportunity for the development of new strategies of analysis. The association tests seek to identify nucleotide variation observed at a genomic region with phenotypic variation for a trait, leading to the identification of DNA regions that control the phenotypes of interest. These tests do not rely on segregating populations, but rather on the evaluation of allelic variation observed in the prospective loci in relation to the phenotype of interest in natural populations. In view of the developments of genomic technology, allelic variations are currently detected by DNA sequencing, characterized as mutations and small insertions and/or deletions (indels) known as SNPs (Collins, Guyer and Chakravarti, 1997).

The rice genome was sequenced in its entirety in 2002 (Goff *et al.*, 2002; Yu *et al.*, 2002), with a great part of the sequences deposited in public databases over the last few months. It is logical to assume that the use of bioinformatics and the re-sequencing of a great number of genes located on some rice chromosomes can be an effective way to correlate the allelic variability detected at these loci with the phenotypic variability for resistance to abiotic stresses. Therefore, re-sequencing of candidate genes in a set of germplasm accessions and phenotypic evaluation of the same set for drought tolerance could be an interesting approach to discover genes involved in the control of traits of economic importance in grass species. Candidate genes found in regions in the vicinity of QTLs for response to the abiotic stresses are immediate targets for drought tolerance gene isolation in this species. Genotypic and phenotypic analysis can be carried out with accessions sampled in germplasm collections, especially landraces of upland rice showing genetic variability for drought tolerance. For this purpose, the following steps should be considered:

Genetic maps making ample use of molecular markers (especially microsatellites) are constructed in order to locate genomic regions (QTLs) associated with the control of drought tolerance in grasses. Several populations of grass species (rice, maize and sorghum) segregating for alleles associated with the control of drought tolerance are then examined. Microsatellite markers are used to cover the entire rice genome by means of large-scale genotyping and mapping. Genotyping is performed using high throughput equipment such as automatic DNA sequencers. By multiplexing many polymerase chain reaction (PCR) products into a single lane on a polyacrylamide gel, a large increase in genotyping throughput is achieved. Multiplexed PCRs greatly increase the amount of collected information in a single reaction. The segregating populations are then submitted to extensive phenotypic evaluation through different bioassays and replicated experiments in the field. The genotypic and phenotypic information is used to identify the genomic regions that control the traits of interest (QTL). The QTL x environment interactions are estimated using the data obtained in different locations. In addition, anchor-markers used for mapping purposes on different grass species are employed in the construction of genetic maps of rice and other grass species, allowing the information obtained in the rice genome to be transferred to these other species via syntenic map analysis, and vice-versa (Moore *et al.*, 1995). Synteny analysis therefore provides a comparative look at the genetic control of drought tolerance in the genome of different species.

The genomic regions delimited by QTL analysis are studied in detail using the rice genome sequence database. The microsatellite markers that flank a specific drought tolerance QTL in a chromosome are used to delimit in the physical map the BAC clones that potentially harbour the gene(s) associated with the trait. All open reading frames (ORFs) and controlling segments in the region are re-annotated and candidate genes are selected for detailed sequence analysis. Genotypic variation (SNPs) are detected after sequence alignment.

The candidate genes selected in the vicinity of a QTL are re-sequenced in a sample of rice variety accessions selected from the gene bank. This sample represents a core collection (Frankel and Brown, 1984) of the germplasm studied, i.e. it should represent the maximum possible level of the genetic diversity existing in a complete germplasm bank.

In-depth phenotypic evaluation of a core collection is undertaken to access diversity of the agronomic trait (drought tolerance) present in the upland rice gene bank. The measurement of drought tolerance is based on indirect observations of morphological, physiological, biochemical and agronomical responses of the plant variety to abiotic stress. Root morphology, for example, is usually dissected in many ways to provide clues on the ability of the plant to survive and compete under drought conditions. In this case, the root system architecture, as measured by the number and arrangement of secondary roots, dry matter weight, root abundance in different layers of soil, and root penetration ability, *inter alia*, are quantified as an indirect way to access drought tolerance. These measurements can be statistically treated as components of drought tolerance. It is also very common to use grain yield under drought conditions to compare genotypic reponses of different plant varieties to water-deficit stresses. Enzyme activity and metabolic bioassays can also offer biochemical means of studying drought tolerance in laboratory conditions.

It is expected that significant associations between sequence polymorphism and phenotypic variation will indicate and allow the isolation of genes that control drought tolerance in rice. This information could be readily transferred to other grass species such as maize and sorghum. The breeding programmes will certainly benefit from the data based on the molecular analysis of gene banks, stimulating an increase in use of stored genetic resources. The integration of genomic technology and characterization of germplasm banks will play an important role in the sustainable conservation of gene banks.

11.4 REFERENCES

Brondani, C., Rangel, P.H.N. & Ferreira, M.E. 2002. QTL mapping and introgression of yield-related traits from *Oryza glumaepatula* to *O. sativa* using microsatellite markers. *Theor. and Appl. Genet.*, 104: 1192-1203.

Collins, F.S., Guyer, M.S. & Chakravarti, A. 1997. Variations on a theme: cataloging human DNA sequence variation. *Science*, 278: 1580-1581.

Ferreira, M.E. & Grattapaglia, D. 1996. *Introdução ao uso de marcadores moleculares em análise genética.* Brasilia, Brazil, Embrapa-SPI. 220 pp.

Frankel, O.H. & Brown, A.H.D. 1984. Current plant genetic resources - a critical appraisal. In *Genetics: new frontiers*, 4: 1-11. New Delhi, India, Oxford & IBH Publishing Co.

Goff, SA, Ricke, D., Lan, T.H., Presting, G., Wang, R., Dunn, M., Glazebrook, J., Sessions, A., Oeller, P., Varma, H. *et al.* 2002. A draft sequence of the rice genome (*Oryza sativa* L. ssp. *japonica*). *Science*, 296: 92-100.

Moore, G., Devos, K.M., Wang, Z.M. & Gale, M.D. 1995. Grasses, line up and form a circle. *Curr. Biol.*, 5: 737-739.

Risch, N.J. 2000. Searching for genetic determinants in the new millennium. *Nature*, 405(15):847-856.

Tanksley, S.D., Ganal, M.W. & Martin, G.B. 1995. Chromosome landing: a paradigm for map-based gene cloning in plants with large genomes. *Trends in Genetics*, 11(2):63-68.

Tanksley, S.D. & Nelson, J.C. 1996. Advanced backcross QTL analysis: a method for the simultaneous discovery and transfer of valuable QTLs from unadapted germplasm into elite breeding lines. *Theor. Appl. Genet.*, 92:191-203.

Tautz, D. 1989. Hypervariability of simple sequences of a general source for polymorphic DNA markers. *Nucleic Acids Research*, 17:6463-6471.

Weber, R.D. & May, P.E. 1989. Abundant class of human DNA polymorphisms which can be typed using the polymerase chain reaction. *American Journal of Human Genetics*, 44: 388-396.

Wehrhahn, C. & Allard, R.W. 1965. The detection and measurement of the effects of individual genes involved in the inheritance of a quantitative character in wheat. *Genetics*, 51:109-119.

Xiao, J., Grandillo, S., Ahn, S.N., McCouch, S.R., Tanksley, S.D., Li, J. & Yuan, L. 1996. Genes from wild rice improve yield. *Nature*, 384: 223-224.

Yu, J., Hu, S., Wang, J., Wong, G.K., Li, S., Liu, B., Deng, Y., Dai, L., Zhou, Y., Zhang, X. *et al.* 2002. A draft sequence of the rice genome (*Oryza sativa* L. ssp. *indica*). *Science*, 296: 79-92.

12. Genetic characterization and its use in decision-making for the conservation of crop germplasm

M. Carmen de Vicente, Felix Alberto Guzmán, Jan Engels and V. Ramanatha Rao

12.1 SUMMARY

This chapter first presents a brief update on the progress made in using genetic characterization to guide decision-making for several conservation activities. It will then focus on the attractive prospects offered by molecular characterization to enhance germplasm use, which is the ultimate purpose of conserving diversity of genetic resources.

12.2 INTRODUCTION

Conservation of genetic resources entails several activities, many of which may greatly benefit from knowledge generated through applying molecular marker technologies. The same applies to activities related to the acquisition of germplasm (locating and describing the diversity), its conservation (using effective procedures) and its evaluation for useful traits. The availability of sound genetic information ensures that decisions made on conservation will be better informed and result in improved germplasm management. Of the activities related to genetic resources, those involving germplasm evaluation and the addition of value to genetic resources are particularly important because they help identify genes and traits, and thus provide the foundation on which to enhance the use of collections.

"Characterization" is the description of a character or quality of an individual (Merriam-Webster, 1991). The word "characterize" is also a synonym of "distinguish", that is, to mark as separate or different, or to separate into kinds, classes or categories. Thus, characterization of genetic resources refers to the process by which accessions are identified or differentiated. This identification may, in broad terms, refer to any difference in the appearance or make-up of an accession. In the agreed terminology of gene banks and germplasm management, the term "characterization" stands for the description of characters that are usually highly heritable, easily seen by the eye and equally expressed in all environments (IPGRI/CIP, 2003). In genetic terms, characterization refers to the detection of variation as a result of differences in either DNA sequences or specific genes or modifying factors. This genetic connotation will be used in this chapter.

Standard characterization and evaluation of accessions may be routinely carried out by using different methods, including traditional practices such as the use of descriptor lists of morphological characters. They may also involve evaluation of agronomic performance under various environmental conditions. In contrast, genetic characterization refers to the description of attributes that follow a Mendelian inheritance or that involve specific DNA sequences. In this context, the application of biochemical assays such as those that detect differences between isozymes or protein profiles, the application of molecular markers and the identification of particular sequences through diverse genomic approaches all qualify as genetic characterization methods.

Because of its nature, genetic characterization clearly offers an enhanced power for detecting diversity (including genotypes and genes) that exceeds that of traditional methods. In addition, genetic characterization with molecular technologies offers greater power of detection than do phenotypic methods (e.g. isozymes). This is because molecular methods reveal differences in genotypes, that is, in the ultimate level of variation embodied by the DNA sequences of an individual and uninfluenced by environment. In contrast, differences revealed by phenotypic approaches are at the level of gene expression (proteins).

12.3 USING MOLECULAR CHARACTERIZATION TO MAKE INFORMED DECISIONS ON THE CONSERVATION OF CROP GENETIC RESOURCES

Information about the genetic make-up of accessions contributes towards decision-making for conservation activities, which range from collecting, managing and identifying genes to adding value to genetic resources.

Well-informed sampling strategies for germplasm material destined for *ex situ* conservation and designation of priority sites (i.e. identifying specific areas with desirable genetic diversity) for *in situ* conservation are both crucial for successful conservation efforts. In turn, defining strategies depends on knowledge of location, distribution and extent of genetic diversity. Molecular characterization, by itself or in conjunction with other data (phenotypic traits or geo-referenced data), provides reliable information for assessing, among other factors, the amount of genetic diversity (Perera *et al.*, 2000), the structure of diversity in samples and populations (Shim and Jørgensen, 2000; Figliuolo and Perrino, 2004), rates of genetic divergence among populations (Maestri *et al.*, 2002) and the distribution of diversity in populations found in different locations (Ferguson, Bramel and Chandra, 2004; Perera *et al.*, 2000).

A recent study on the genetic diversity of cultivated *Capsicum* species in Guatemalan home gardens compared the diversity present in an array of home gardens in the Department of Alta Verapaz with a countrywide representative sample of 40 accessions conserved *ex situ* in the national collection (Guzmán *et al.*, 2005). The results showed that home gardens of Alta Verapaz (H = 0.251) contained as much diversity as the entire national *ex situ* collection (H = 0.281). These results thus suggest that (i) home gardens are indeed an extremely important resource for *in situ* conservation of *Capsicum* germplasm in Guatemala, and as such should not be neglected; (ii) if further collecting activities were to be undertaken, special emphasis should be given to collecting in Alta Verapaz; and (iii) additional collecting in Alta Verapaz alone could disclose novel genetic diversity that is absent from the national collection.

Conservation of clonally propagated crops demands more complex and expensive procedures. If these crops are maintained on-farm, their existence is endangered by several factors, including the introduction of alternative improved varieties. Conservation efforts thus need to be based on solid knowledge of clonal diversity. This was the case for Abyssinian banana, or ensete (*Ensete ventricosum* [Welw.] Cheesman) from Ethiopia, which was analysed with amplified fragment length polymorphism (AFLP) markers (Negash *et al.*, 2002). Of the 146 clones from five different regions, only 4.8 percent of the total genetic variation was found between regions, whereas 95.2 percent was found within regions. The results led to a reduced number of clones for conservation and indicated the existence of a common practice of exchange of local types between regions, which, in its turn, emphasized the need to collect further in different farming systems.

A study on taro (*Colocasia esculenta* [L.] Schott) genetic diversity in the Pacific, using simple sequence repeats (SSR) markers, showed that many of the accessions from countries of the Pacific region were identical to those of Papua New Guinea. This indicates that originally the cultivars may have been introduced throughout the region from Papua New Guinea (Mace *et al.*, 2005) and that collection of taro genetic diversity could focus on Papua New Guinea alone.

Molecular characterization also helps determine the breeding behaviour of species, individual reproductive success and the existence of gene flow, that is, the movement of alleles within and between populations of the same or related species, and its consequences (Papa and Gepts, 2003). Molecular data improve or even allow the elucidation of phylogeny, and provide the basic knowledge for understanding taxonomy, domestication and evolution (Nwakanma *et al.*, 2003). As a result, information from molecular markers or DNA sequences offers a good basis for better conservation approaches.

Management of germplasm established in a collection, usually a field, seed or *in vitro* gene bank, comprises several activities. These activities usually seek to ensure the identity of the individually stored and maintained samples, the safeguarding of genetic integrity and genetic diversity, and material available for distribution to users. These tasks, which are primarily the responsibility of gene bank managers and curators, involve the control of accessions on arrival at the facilities and their continuous safeguarding for the future through regeneration and multiplication. For all these routine activities, information about the genetic constitution of samples or accessions is critical and possibly provides the most important means of measuring the quality of the work being performed.

Börner, Chebotar and Korzun (2000) analysed bulk seed of wheat accessions to test their genetic integrity after 24 cycles of regeneration and after more than 50 years of storage at room temperature in a gene bank. They found neither contamination nor incorrect manipulation effects such as mechanical mixtures, but did identify one case of genetic drift in one accession. By splitting its germplasm samples into either almost or completely pure lines, i.e. accessions, the IPK-Gatersleben gene bank (Germany) is expected to have contributed to this very positive finding (J. Engels, personal communication, 2005).

In the same gene bank, a study examined the genetic constitution of rye accessions that underwent frequent regeneration. Results showed that a significant number of alleles present in the original sample was lacking in the newly regenerated material and new alleles in the new material were not present in the first regeneration sample (Chebotar *et al.*, 2003). Thus, the use of molecular markers can quickly help check whether changes in alleles or allele frequencies are taking place.

Molecular information has been used to weigh the need for decreasing the size of germplasm collections, which otherwise would add costs to the long-term conservation of germplasm. For instance, Dean *et al.* (1999) used microsatellite markers to analyse the genetic diversity and structure of 19 sorghum accessions known as "Orange" in the national sorghum collection of the United States Department of Agriculture (USDA). They found two redundant groups (involving five entries) among the 19 accessions evaluated. They also found that much of the total genetic variation was partitioned among accessions. As a result, the authors concluded that the number of accessions held by the US National Plant Germplasm System (NPGS) could be significantly reduced without risking the overall amount of genetic variation contained in these holdings.

Markers were also helpful in examining genetic identities and relationships of *Malus* accessions (Hokanson *et al.*, 1998). Eight primer pairs unambiguously differentiated 52 of 66 genotypes in a study that calculated the probability of any two genotypes being similar at all loci analysed as being about one in a thousand million. The results not only discriminated among the genotypes, but were also shown to be useful for designing strategies for the collection and *in situ* conservation of wild *Malus* species.

Selected molecular technologies render cost-effective and comprehensive genotypic profiles of accessions ("fingerprints") that may be used to establish the identity of the material under study. Simultaneously, in addition to the presence of redundant materials (or "duplicates") (McGregor *et al.*, 2002), these technologies can detect contaminants, and in the case of material mixtures, contamination with introgressed genes from other accessions or commercial varieties as well. Moreover, molecular data provide the baseline for monitoring natural changes in the genetic structure of the accession (Chwedorzewska, Bednarek and Puchalski, 2002), or those occurring as a result of human intervention (e.g. seed regeneration or sampling for replanting in the field). Whatever the case, analysis of molecular information allows the design of strategies for either purging the consequences of inappropriate procedures or amending them to prevent future inconveniences (de Vicente, 2002).

A small number of potential duplicates were identified in a core collection of cassava (Manihot esculenta Crantz) when isozyme and AFLP profiles were compared (Chavarriaga-Aguirre *et al.*, 1999). The core collection had been assembled with information from traditional markers, which proved to be highly effective for selecting unique genotypes. Molecular data were used for efficiently verifying the previous work on the collection and ensure minimum repetition. The taro core collection for the Pacific region was treated in a similar manner (Mace *et al.*, 2005). Thus, gene bank managers can easily realize the potential value of using molecular methods to support and possibly modify or improve the operations of a gene bank.

A special and increasingly important role of genetic characterization is identifying useful genes in germplasm, that is, maximizing conservation efforts. Because the major justification for the existence of germplasm collections is for the use of the conserved accessions, it is important to identify the valuable genes that can help develop varieties that will be able to meet the challenges of current and future agriculture.

Characterization has benefited from several approaches resulting from advances in molecular genetics such as genetic and quantitative trait locus (QTL) mapping, and gene tagging (Yamada *et al.*, 2004; Kelly *et al.*, 2003). Research in this field has led to the acknowledgement of the value of wild relatives, in which modern techniques have discovered useful variation that could contribute to varietal improvement (Xiao *et al.*, 1996; de Vicente and Tanksley, 1993). Knowledge of molecular information in major crops and species and of the synteny of genomes, especially conservation of gene order, has also opened up prospects for identifying important genes or variants in other crop types, particularly those that receive little attention from formal research.

12.4 FUTURE TRENDS

Most marker technologies target genomic regions, which are selectively neutral; some technologies, however, target specific genes. The neutrality of markers is suitable for most uses in germplasm conservation and management. However, when the interest of conservation lies specifically in the diversity of traits of agronomic importance, some questions remain on the markers' representativeness. In such cases, the markers able to detect functional diversity are more suitable for characterizing germplasm collections.

Germplasm in collections can undergo structural molecular characterization, i.e. based on the building blocks of the DNA sequence, and functional molecular characterization, i.e. based on the identification of genes and their functions. Such characterization permits access to the raw materials - the genes - for nearly all the objectives of today's and tomorrow's breeding programmes. The information gathered from structural characterization not only provides increased clarity on existing genetic diversity and its organization in individuals, but also determines sample and population organization that may ultimately form the basis for functional characterization.

The increasing number of sequencing projects has resulted in an increased opportunity to produce expressed sequence tags (ESTs) to which gene functions may be assigned. Moreover, such projects allow for the compilation of an enormous amount of sequence data that can be used to develop markers linked to specific genes, which in turn may help identify novel functional variation (Han *et al.*, 2004; NCBI, 2001).

In addition, the development of novel technologies continues. This usually means decreased costs - a very significant point for their application in the tasks of conserving genetic resources, which tend to involve large numbers of samples and to have difficulties in sourcing needed funds. Other improvements involve increasing the throughput, both in number of markers analysed and in number of samples, and simplifying technologies.

New developments are also taking place in designing better approaches to access new and useful genetic variation in collections, namely, allele mining and association genetics. Allele mining focuses on the detection of allelic variation in important genes and/or traits within a germplasm collection (Simko *et al.*, 2004b). If the targeted DNA (either a gene of known function or a given sequence) is known, then the allelic variation (usually point mutations) in a collection can be identified using methods developed for the purpose (Lemieux, Aharoni and Schena, 1998).

Association studies of artificial progenies are an alternative to segregation analysis for identifying useful genes by correlation of molecular markers and a specific phenotype (Gebhardt *et al.*, 2004). Association studies can be performed on a germplasm collection and also on other materials as long as significant linkage disequilibrium (LD) exists, for example, breeding materials. It may be especially useful for those crops where appropriate populations for genetic analysis cannot be obtained or their production is too time-consuming (Simko *et al.*, 2004a). It is also useful for those crops for which sequence information does not exist and is unlikely to be available soon.

12.5 THE CHALLENGES AHEAD

The importance of the variation captured in genetic resources in allowing evolution and/or facilitating plant breeding has been long recognized. However, appreciating the variation held in collections is not sufficient. Conservation of genetic resources needs to go hand in hand with enhanced use of the conserved material. Identifying and making available the allelic variation that makes up the genotype and phenotype provide the groundwork on which genetic resources can be used in, for example, plant breeding.

The number of accessions held collectively by all Consultative Group on International Agricultural Research (CGIAR) gene banks is estimated at almost 600 000 (FAO, 1998). Together with the collections established by national programmes worldwide, this number reaches almost 6 million (FAO, 1998). Without doubt, these genetic resources collections, together with uncollected germplasm and that held *in situ* and on-farm, harbour abundant quantities of hidden allelic variants. The challenge is to unravel the mysteries of this variation so that it can be used for the benefit of humankind.

Gene banks hold large numbers of accessions, particularly of staple crops. Modern improvements in equipment and procedures allow considerable sample throughput. This can be costly. However, the more a technology develops, the lower its costs will be per data point and per sample. Nevertheless, the higher the throughput used, the higher the number of data points obtained. This requires adequate equipment for handling and storing and expertise for handling and analysing in order to draw adequate results from the investment.

One possible avenue for ensuring broader benefits from molecular characterization is the establishment of international collaboration for particular crops. Although equipment and expertise cannot currently be readily available worldwide, characterization networks are possible. In addition to carrying out the laboratory work, such networks would also facilitate access to information, thus fostering closer links between curators, breeders and molecular scientists. At the same time, countries with little expertise or equipment can make steady progress in both areas, thus making better use of the genetic resources that they hold (Hamon, Frison and Navarro, 2004).

More and more, technologies have increased throughputs, which generally means the generation of progressively larger amounts of data. Such data should not languish unused. If gene banks equip themselves with the latest technologies, then they should be able to translate such data into scientific knowledge. To do so, they need not only laboratory technical expertise, but also bioinformatics staff. This means that through their molecular work, gene banks may keep not only live plant materials, but also DNA and data. Hence, the banks may develop appropriate new roles as providers of genetic resources and their accompanying data in an array of forms. Such broadening of the gene banks' roles implies that their clientele will also expand from plant breeders to include molecular geneticists, molecular biologists and even bioinformaticists. The expanded range of roles may even lead to including activities related to phenotyping, a type of characterization beyond the traditional description of morphology and general field performance. Phenotyping is very much linked to the usefulness of good molecular characterization, together forming the basis of progress in modern genomics research (de Vicente, 2004).

12.6 CONCLUSIONS

The most important challenges in the near future are certainly the identification of useful variation (real or potential) in germplasm and its use in guiding conservation decisions. Knowing the presence of useful genes and alleles would help in making decisions on the multiplication of accessions and the maintenance of seed stocks when responding to an expected higher demand for materials. Such information may also help in making decisions on heterogeneous accessions where only some genotypes may possess useful alleles. The gene bank curator may have to decide on maintaining the original material as is and separating a subpopulation carrying the desirable alleles as well as giving it new accession numbers and management protocols. This will facilitate germplasm use and add value to the collections.

Similarly, genotypes with known and interesting genes and alleles can be added to core collections to make them more useful to the user community. To promote use of the main collection, a core collection is developed to capture 75 to 80 percent of the representative genetic diversity. From the user's perspective, such a core collection will gain value when accessions with known genes are added, even if the general genetic diversity present in these few additional accessions is already present in the core collection.

Finally, an extreme concept that could arise, based on the knowledge of the presence of valuable genes and alleles, is that of building collections based on traits. This is not a novel idea *per se*, but the initiative may come about once sufficient genomic results become available. Certainly, recent scientific advancements are drawing closer to this future.

12.7 ACKNOWLEDGEMENTS

The authors would like to thank Ehsan Dulloo (IPGRI, headquarters) for his critical reading of the chapter, Dimary Libreros (IPGRI Office for the Americas) for her bibliographic support and Elizabeth McAdam for copy editing.

12.8 REFERENCES

Börner, A., Chebotar S. & Korzun V. 2000. Molecular characterization of the genetic integrity of wheat (*Triticum aestivum* L.) germplasm after long-term maintenance. *Theor. Appl. Genet.*, 100: 494-497.

Chavarriaga-Aguirre, P., Maya M.M., Tohme J., Duque M.C., Iglesias C., Bonierbale M.W., Kresovich S. & Kochert G. 1999. Using microsatellites, isozymes and AFLPs to evaluate genetic diversity and redundancy in the cassava core collection and to assess the usefulness of DNA-based markers to maintain germplasm collections. *Mol. Breeding*, 5: 263-273.

Chebotar, S., Roder, M.S., Korzun, V., Saal, B., Weber, W.E. & Börner, A. 2003. Molecular studies on genetic integrity of open-pollinating species rye (*Secale cereale* L.) after long-term genebank maintenance. *Theor. Appl. Genet.*, 107:1469-1476.

Chwedorzewska, K.J., Bednarek P.T. & Puchalski J. 2002. Studies on changes in specific rye genome regions due to seed aging and regeneration. *Cell. Mol. Biol. Lett.*, 7: 569-576.

Dean, R.E., Dahlberg, J.A., Hopkins, M.S., Mitchell, S.E. & Kresovich, S. 1999. Genetic redundancy and diversity among 'Orange' accessions in the U.S. national sorghum collection as assessed with simple sequence repeat (SSR) markers. *Crop Sci.*, 39: 1215-1221.

de Vicente, M.C. & Tanksley S.D. 1993. QTL analysis of transgressive segregation in an interspecific tomato cross. *Genetics*, 134(2): 585-596.

de Vicente, M.C. 2002. Molecular techniques to facilitate prioritization of plant genetic resources conservation and further research. *AgBiotechNet*, 4. ABN 092.

de Vicente, M.C. 2004. *The evolving role of genebanks in the fast-developing field of molecular genetics*, Introduction, pp. 7-12. *In* M.C. de Vicente, ed. *Issues in genetic* *resources, No.* XI, August 2004. Rome, International Plant Genetic Resources Institute.

FAO. 1998. The state of the world's plant genetic resources for food and agriculture. Rome.

Ferguson, M.E., Bramel, P.J. & Chandra, S. 2004. Gene diversity among botanical varieties in peanut (*Arachis hypogaea* L.). *Crop Sci.*, 44: 1847-1854.

Figliuolo, G. & Perrino, P. 2004. Genetic diversity and intra-specific phylogeny of *Triticum turgidum* L. subsp. *dicoccon* (Schrank) Thell. revealed by RFLPs and SSRs. *Genet. Resour. Crop Ev.*, 51: 519-527.

Gebhardt, C., Ballvora, A., Walkemeier, B., Oberhagemann, P. & Schüler, K. 2004. Assessing genetic potential in germplasm collections of crop plants by marker-trait association: a case study for potatoes with quantitative variation of resistance to late blight and maturity type. *Mol. Breeding* 13: 93-102.

Guzmán, F.A., Ayala H., Azurdia C., Duque, M.C. & de Vicente, M.C. 2005. AFLP assessment of genetic diversity of *Capsicum* genetic resources in Guatemala: home gardens as an option for conservation. *Crop Sci.*, 45: 363-370.

Hamon, S., Frison, E. & Navarro, L. 2004. Connecting plant germplasm collection and genomic centres: How to better link curators, molecular biologists and geneticists? pp. 33-42 *in* M.C. de Vicente, ed. *The evolving role of gene banks in the fast-developing field of molecular genetics. Issues in genetic resources*, No. XI, August 2004. Rome, International Plant Genetic Resources Institute.

Han, Z.G., Guo, W.Z., Song, X.L. & Zhang, T.Z. 2004. Genetic mapping of EST-derived microsatellites from the diploid *Gossypium arboretum* in allotetraploid cotton. *Mol. Gen. Genomics*, 272: 308-327.

Hokanson, S.C., Szewc-McFadden, A.K., Lamboy, W.F. & McFerson, J.R. 1998. Microsatellite (SSR) markers reveal genetic identities, genetic diversity and relationships in a *Malus x domestica* borkh. core subset collection. *Theor. Appl. Genet.*, 97: 671-683.

International Plant Genetic Resources Institute (IPGRI)/Centro Internacional de la Papa (CIP). 2003. Descriptores del Ulluco (*Ullucus tuberosus*). Rome, IPGRI; Lima, Peru, CIP.

Kelly, J.D., Gepts, P., Miklas, P.N. & Coyne, D.P. 2003. Tagging and mapping of genes and QTL and molecular marker-assisted selection for traits of economic importance in bean and cowpea. *Field Crop Res*, 82(2/3): 135-154.

Lemieux, B., Aharoni, A. & Schena, M. 1998. Overview of DNA chip technology. *Mol. Breeding* 4: 277-289.

Mace, E.S., Mathur P.N., Godwin, I.D., Hunter, D. Taylor, M.B., Singh, D., DeLacy, I.H. & Jackson, G.V.H. 2005. Development of a regional Core Collection (Oceania) for taro, *Colocasia esculenta* (L.) Schott., based on molecular and phenotypic characterization. (in press). *In* V. Ramanatha Rao, P.J. Matthews & P.B. Eyzaguirre, eds. *The global diversity of taro: ethnobotany and conservation.* Rome, IPGRI; Osaka, Japan, Minpaku (National Museum of Ethnology, Osaka, Japan).

Maestri, E., Malcevschi A., Massari, A. & Marmiroli N. 2002. Genomic analysis of cultivated barley (*Hordeum vulgare*) using sequence-tagged molecular markers. Estimates of divergence based on RFLP and PCR markers derived from stress-responsive genes, and simple-sequence repeats (SSRs). *Mol. Gen. Genomics*, 267(2): 186-201.

McGregor, C.E., van Treuren, R., Hoekstra, R. & van Hintum, Th.J.L. 2002. Analysis of the wild potato germplasm of the series Acauliawith AFLPs: implications for *ex situ* conservation. *Theor. Appl. Genet.*, 104: 146-156.

Merriam-Webster. 1991. *Webster's ninth new collegiate dictionary.* Springfield, Massachusetts, USA, Merriam-Webster Inc. Publishers.

National Center for Biotechnology Information (NCBI). 2001. *ESTs: gene discovery made easier.* (available at www.ncbi.nlm.nih.gov/About/primer/est.html)

Negash, A., Tsegaye A., van Treuren, R. & Visser, B. 2002. AFLP Analysis of enset clonal diversity in South and Southwestern Ethiopia for conservation. *Crop Sci.*, 42: 1105-1111.

Nwakanma, D.C., Pillay, M., Okoli, B.E. & Tenkouano, A. 2003. Sectional relationships in the genus *Musa* L. inferred from the PCR-RFLP of organelle DNA sequences. *Theor. Appl. Genet.*, 107: 850-856.

Papa, R. & Gepts, P. 2003. Asymmetry of gene flow and differential geographical structure of molecular diversity in wild and domesticated common bean (*Phaseolus vulgaris* L.) from Mesoamerica. *Theor. Appl. Genet.*, 106: 239-250.

Perera, L., Russell J.R., Provan, J. & Powell, W. 2000. Use of microsatellite DNA markers to investigate the level of genetic diversity and population genetic structure of coconut (*Cocos nucifera* L.). *Genome* 43: 15-21.

Shim, S.I. & Jørgensen, R.B. 2000. Genetic structure in cultivated and wild carrots (*Daucus carota* L.) revealed by AFLP analysis. *Theor. Appl. Genet.*, 101: 227-233.

Simko, I., Costanzo, S., Haynes, K.G., Christ, B.J. & Jones, R.W. 2004a. Linkage disequilibrium mapping of a *Verticillium dahliae* resistance quantitative trait locus in tetraploid potato (*Solanum tuberosum*) through a candidate gene approach. *Theor. Appl. Genet.*, 108: 217-224.

Simko, I., Haynes, K.G., Ewing, E.E., Costanzo, S., Christ, B.J. & Jones, R.W. 2004b. Mapping genes for resistance to *Verticillium albo-atrum* in tetraploid and diploid potato populations using haplotype association tests and genetic linkage analysis. *Mol. Gen. Genomics*, 271: 522-531.

Yamada, T., Jones E.S., Cogan N.O.I., Vecchies A.C., Nomura T., Hisano H., Shimamoto Y., Smith, K.F., Hayward, M.D. & Forster, J.W. 2004. QTL analysis of morphological, developmental, and winter hardiness-associated traits in perennial ryegrass. *Crop Sci.*, 44: 925-935.

Xiao, J., Grandillo, S., Ahn, S.N., Mccouch, S.R., Tanksley, S.D., Li, J.M. & Yuan, L.P. 1996. Genes from wild rice improve yield. *Nature* (London), 384 (6606): 223-224.

13. The role of biotechnology in the conservation, sustainable use and genetic enhancement of bioresources in fragile ecosystems

Prashanth S. Raghavan and Ajay Parida

13.1 SUMMARY

Mangroves are plants with the ability to tolerate a high degree of soil salinity and various other forms of abiotic stress. They are woody and generally inhabit the upper intertidal zones of estuaries within the tropical and subtropical regions. Anthropogenic influences have led to a significant reduction in its distribution worldwide and local extinction of many species and populations of the mangrove ecosystem. Our group therefore initiated a study on the molecular biology of mangroves, aiming to help in the conservation of these species. The study on mangroves in our laboratory began by analysing the genetic diversity in the various species present in the mangrove species. Twenty-four species of mangroves and mangrove associates were analysed using molecular markers. This study resulted in species-specific restriction patterns in the genera *Rhizophora* and *Sueda*, observing intra-generic variations in three genera - *Avicennia*, *Rhizophora* and *Sueda*. Intra/interspecific variation in the genus *Avicennia* and analysis of mitochondrial DNA variation in the species of Rhizophoraceae were also studied. As a first step towards characterizing genes that contribute to combating salinity stress, we constructed complementary DNA (cDNA) libraries from a mangrove species. An analysis of the expressed sequence tags (ESTs) generated from the cDNA library of one mangrove species revealed that 30 percent of the analysed ESTs showed homology to previously uncharacterized genes in the public plant databases. Several full-length genes of proteins involved in salinity tolerance and antioxidative stress pathways were isolated and submitted to the National Centre for Biotechnology Information (NCBI) gene bank database. Some of these isolated genes from the mangrove have been transferred to rice and homozygous lines of these transgenic rice plants have been generated, which are being tested for tolerance to various abiotic stresses at the laboratory and field level.

13.2 INTRODUCTION

The coastal ecosystem is one of the most productive ecosystems. The tropical and the subtropical coastlines of the world are characterized by specialized littoral plant formations, mangroves (Lakshmi, Parani and Parida, 2001). A mangrove is a plant community that inhabits the boundary between terrestrial and aquatic environments, comprising around 71 species, spread over 16 families (ibid.). All the mangrove species have the same physiognomy, physiological characteristics and structural adaptations (Yenny-Esinsine, 1980). The plant species in this particular ecosystem are constantly under varied environmental stress conditions including high saline and temperature extremes; these plants have adapted themselves to these frequent and fluctuating changes (Lakshmi *et al.*, 1997).

However, human interference and negligence have caused rapid destruction to the mangrove forests and even caused extinction to a few of the locally existing species (Parani *et al.*, 2000). Large areas of mangrove forests throughout the world are being converted for agriculture or exploited for wood and other forest products (Lakshmi *et al.*, 1997). The Indian coastline covers around 7 500 km and accounts for 8 percent of the world's mangrove area, and the eastern coast of India accounts for about 82 percent of the mangrove forest cover in India (Parida *et al.*, 1998). In the absence of any national plan for conservation and sustainable utilization, mangroves along the Indian coast have reached an alarming stage of depletion (Lakshmi *et al.*, 1997). A 25 percent reduction in mangrove forest cover has been reported along the Indian region during the last 25 years (Parida *et al.*, 1998).

In the coastal regions there is intense agricultural activity. Increased soil erosion and water pollution caused by intensive farm practices in the inland area gets transported through the rivers and canals and adversely affects the coastal agro-ecosystem. The seawater intrusion and the attendant soil and water quality problems caused by the groundwater depletion have already started threatening the sustainability of the agricultural ecosystem in the Saurashtra region of Gujarat and Thanjavur region of Tamil Nadu in India. This has given rise to an increase in the level of abiotic stresses such as salinity, alkalinity and drought. Climate change with its consequent rise in sea level is one of the major impending dangers affecting the coastal ecosystem. By 2025, the rise in sea level is expected to be roughly 8 to 29 cm due to global warming (Parida *et al.*, 1998). This could cause large-scale inland flooding.

Salinization is posing an increasing problem in coastal and agricultural areas, reducing plant productivity and yield. Further, salinity is one of the major abiotic stresses decreasing plant productivity. Tolerance to salt stress is a complex trait that involves various aspects including osmotic, ionic stress and secondary stress, such as oxidative stress. Salt stress leads to dehydration and osmotic stress, with the reduced availability of water resulting in stomatal closure, and reduced supply of carbon dioxide leading to a high production of reactive oxygen species in the chloroplasts (Tanaka *et al.*, 1999). This effect causes irreversible cellular damage. Similar effects of generation of reactive oxygen species are also seen to occur during periods of high photosynthetic activity in plants, resulting in photoinhibition (Bowler, Montagu and Inze, 1992). Photoinhibition and salinity stress together cause severe damage to the cellular processes in the plant. However, mangroves are plants that are capable of surviving in highly saline environments, having a high capacity to maintain active leaves in conditions that are expected to severely reduce the photosynthesis through photoinhibition (Cheeseman *et al.*, 1997). In order to combat such abiotic stress effects, we have carried out studies to conserve the mangrove genetic resources, characterize and harness the genes involved in salinity/abiotic stress tolerance from mangroves, and transfer these genes to crop plants in order to generate crops with enhanced stress tolerance.

13.3 CYTOLOGICAL AND GENETIC DIVERSITY ANALYSIS OF MANGROVES

The cytological and genetic diversity present in the mangrove plants were studied as an initial part of the work. Mitotic chromosome analysis carried out on samples of different populations of *Acanthus illicifolius* (Lakshmi *et al.*, 1997) revealed that the cells were characterized by 48 chromosomes resolved into 24 homomorphic pairs. There was no variation in the chromosome number seen in the plants from different populations. Chromosome analysis of ten species of the genus Rhizophoraceae (Lakshmi, Parani and Parida, 2001) revealed that there were no numerical variations in the intraspecific level and that the chromosome complements in these species were stable and underwent limited divergence during speciation.

Genetic diversity studies were carried out for intra/interspecific variations in different populations of the genus *Avicennia* (Parani *et al.*, 1997). A mangrove genetic resource centre was established in the Pichavaram mangrove area, Chidambaram, India, where the endangered mangrove species are being conserved. Following the study on the genetic diversity present in *Avicennia*, seeds from all the populations studied were collected to represent *A. marina* in the mangrove genetic resources centre, and the sampling of the seeds was based on the recommendation specified by the outcome of the study. Intraspecific variation in the species *Excoecaria agallocha* was also studied (Lakshmi *et al.*, 2000). The study revealed that lack of morphological variation in the plants belonging to this species was not due to lack of genetic variation.

Subsequently, 24 species of mangroves and mangrove associates were analysed using molecular markers. This study enabled the generation of species-specific restriction patterns in the genera *Rhizophora* and *Sueda*. We were able to study the intrageneric variation in three genera, *Avicennia*, *Rhizophora* and *Sueda* (Parani *et al.*, 2000). Analysis of mitochondrial DNA variation in species of Rhizophoraceae (Lakshmi *et al.*, 2002) was also studied and again resulted in species-specific profiling of different species belonging to the family of Rhizophoraceae.

13.4 ISOLATION OF SALINITY TOLERANCE GENES FROM MANGROVES

As a first step towards characterizing genes from mangroves that contribute to improving salinity stress, we constructed a cDNA library from a mangrove species *Avicennia marina*, using seedlings treated for 48 hours with 0.5M NaCl (Parani *et al.*, 2002). Comparison was made between 1841 ESTs from *A. marina* cDNA library against sequences from the NCBI databases using the program BLASTX. Unknown genes form the largest category at 30 percent, followed by genes required for primary metabolism (13 percent). Genes involved in transcription and chromatin organization, protein synthesis and processing each represent 10 percent of the sequenced ESTs while those involved in membrane transport and intracellular trafficking represent 9 percent of the ESTs. Eight percent of the ESTs relate to signal transduction while 7 percent are similar to previously reported stress-induced genes (Preeti *et al.*, 2005).

13.5 GENETIC ENGINEERING FOR SALINITY TOLERANCE

The mangrove genes isolated from the mangroves were transferred into crop plants using *Agrobacterium tumefaciens* mediated transformation (Hiei *et al.*, 1994). Specific genes isolated from the *A. marina* cDNA library were cloned in binary vectors. These genes were expressed under the control of constitutive promoters. These gene constructs were then transformed into *Agrobacterium tumefaciens* and used for co-cultivation with rice calli and tobacco leaf explants. Rice calli were generated from seed scutella of mature rice seeds on a callus induction medium. Callus co-cultivated with *Agrobaterium* were washed and selected on callus induction medium containing hygromycin. The selected calli were then transferred to the regeneration medium containing shoot-inducing hormones. The regenerants were transferred to the rooting medium. The plantlets were subsequently transferred to a hardening medium. Finally, these plants were transferred to the soil in pots to raise the next generation seeds. The tillers of the rice plants were bagged before the onset of flowering in order to promote self-pollination. The seeds from the selfed plants were collected and again sown for the next generation. In a similar manner, leaf discs from tobacco plants were infected with *Agrobacterium* carrying the binary constructs harbouring the mangrove genes. The integration and expression of the transgenes were confirmed in these plants using various molecular analyses such as PCR, Southern hybridization, Northern hybridization, isozyme analysis and Western blot analysis.

The homozygous lines from these transgenic plants were raised and tested with various abiotic stresses such as salt stress and drought stress. Initial analyses in the laboratory have been promising. However, further analyses would need to be carried out to evaluate the performance of these transgenics. It was found that all the transgenic lines performed better than the control in stress conditions. This effort happens to be the first effort of its kind in the world in terms of transferring genes from *A. marina* to crop plants. This process therefore has a multiple use by conserving the genetic diversity of the mangrove species, which is being destroyed very rapidly, and by genetically engineering the crop plant with genes, which help them tolerate abiotic stress better.

13.6 REFERENCES

Bowler, C., Van Montagu, M. & Inze, D. 1992. Superoxide dismutase and stress tolerance, *Annual review of plant physiology and plant molecular biology,* 43:83-116.

Cheeseman, J.M., Herendeen, L.B., Cheeseman, A.T. & Clough, B.F. 1997. Photosynthesis and photoprotection in mangroves under field conditions. *Plant, Cell and Environment,* 20: 579-590.

Hiei, Y., Ohta, S., Komari,T & Kumashiro, T. 1994. Efficient transformation of rice (*Oryza sativa* L.) mediated by Agrobacterium and sequence analysis of the boundaries of the T-DNA. *The Plant Journal,* 6(2): 271-282.

Lakshmi, M., Parani, M., Nivedita, R. & Parida, A. 2000. Molecular phylogeny of mangroves. VI. Intraspecific genetic variation in mangrove species *Excoecaria agallocha* L. (Euphorbiaceae) *Genome,* 43: 110-115.

Lakshmi, M, Parani, M. & Parida, A. 2001. Genetic diversity in species of India mangrove family Rhizophoraceae. *Forest genetic resources: status, threats and conservation strategies.* New Delhi, India, Oxford IBH Publications. pp 31-47.

Lakshmi, M., Parani, M., Senthilkumar, P. & Parida, A. 2002. Molecular phylogeny of mangroves VIII: Analysis of mitochondrial DNA variation for species identification and relationships in Indian mangrove *Rhizophoraceae. Wetlands ecology and management,* 10: 355-362.

Lakshmi, M., Rajalakshmi, S., Parani, M., Anuratha, C.S. & Parida, A. 1997. Molecular phylogeny of mangroves I. Use of molecular markers in assessing the intraspecific genetic variability in the mangrove species *Acanthus ilicifolius* Linn. (Acanthaceae). *Theoretical and Applied Genetics,* 94:1121-1127.

Parani, M., Jithesh, M.N., Lakshmi, M. & Parida, A. 2002. Cloning and characterization of a gene encoding ubiquitin conjugating enzyme from the mangrove species. *Avicennia marina* (Forsk.) Vierh. *Indian Journal of Biotechnology,* 1: 164-169.

Parani, M., Lakshmi, M., Elango, S., Nivedita, R., Anuratha, C.S & Parida, A. 1997. Molecular phylogeny of Mangroves II. Intra and inter specific variation in *Avicennia* revealed by RAPD and RFLP markers. *Genome,* 40:487-495.

Parani, M., Lakshmi, M., Zeigenhagen, B., Faldung, M., Senthilkumar, P. & Parida, A. 2000. Molecular phylogeny of mangroves VII. PCR-RFLP of trnS-psbC and rbcL gene regions in 24 mangrove and mangrove-associate species. *Theoretical and Applied Genetics,* 100:454-460.

Parida, A., Parani, M., Lakshmi, M., Nivedita, R., Elango, S. & Anuratha, C.S. 1998. Nature and extent of genetic variation and species diversity in Indian mangroves. *IAEA Techdoc,* 1047: 95-105.

Preeti, A.M., Sivaprakash, K.R., Parani, M., Gayatri Venkataraman & Parida, A. 2005. Generation and analysis of expressed sequence tags from the salt-tolerant mangrove species *Avicennia marina* (Forsk) Vierh. *Theoretical and Applied Genetics*, 110: 416-424.

Tanaka, Y., Hibino, T., Hayashi, Y., Tanaka, A., Kishitani, S., Takabe, T., Yokota, S. & Takabe, T. 1999. Salt tolerance of transgenic rice overexpressing yeast mitochondrial Mn-SOD in chloroplasts. *Plant Science*, 148: 131-138.

Yenny-Esinsine. 1980. *Elements of tropical ecology*. London, England, Heinemann Educational Books.

14. Genetic diversity in forest tree populations and conservation: analysis of neutral and adaptive variation

Giovanni G. Vendramin and Michele Morgante

14.1 SUMMARY

Conservation genetics of forest tree species takes advantage of the availability of molecular markers. Depending on the processes that need to be analysed, molecular markers may provide extremely useful information to monitor processes related to adaptation and migration. Markers must be carefully selected depending on the specific question. Some examples are provided and the usefulness of molecular markers in conservation genetics of forest tree species is discussed. Direct analysis of adaptive variation is now possible by analysis of quantitative trait nucleotides (QTNs), which requires that QTNs be known. A perspective on the possibilities for discovering genes involved in adaptive variation in forest trees is provided.

14.2 CONTRIBUTION

The genomics revolution of the last ten years has improved our understanding of the genetic make-up of living organisms. Together with the achievements represented by complete genomic sequences for an increasing number of species, high throughput and parallel approaches are available for the analysis of transcripts, proteins, insertional and chemically-induced mutants. All this information facilitates the understanding of the function of genes in terms of their relationship to the phenotype. Despite its great relevance, such an understanding could be of little value to population and conservation genetics because it will not elucidate the relationship between genetic variation in gene sequences and phenotypic variation in traits, but rather only that between a gene and a mutant phenotype. The relationships between complex trait variation and molecular diversity of genes can be studied based on a genomic approach, but the identification of genes responsible for the variation remains a slow and time-consuming process, especially in long-living organisms such as forest trees. Work in model plant species such as *Arabidopsis* and rice has, however, started to unveil an ever-increasing number of genes involved in the determination of traits of adaptive significance, such as phenology and abiotic stress tolerance/resistance. This progress will finally allow ecological and conservation genetics to directly analyse variation in genes involved in adaptive processes rather than in neutral markers. Neutral markers will, however, remain important to make inferences about stochastic processes affecting natural population evolution.

Populations may follow two main strategies to react to abiotic stresses originating from climate changes: adapt to the new climatic conditions and/or migrate to more favourable areas. Some neutral markers (e.g. highly polymorphic microsatellites and organelle markers) are very useful for monitoring past and present migration processes. Plants offer excellent models to investigate how gene flow shapes the organization of genetic diversity. Their three genomes (chloroplast, mitochondrial and nuclear) can have different modes of transmission and will hence experience varying levels of gene flow. Based on a very large data set, Petit *et al.* (2005) demonstrates that mode of inheritance appears to have a major effect on genetic differentiation (G_{ST}). G_{ST} for chloroplast DNA and mitochondrial DNA markers covary narrowly when both genomes are maternally inherited. At the range-wide scale, historical levels of pollen flow are generally at least an order of magnitude larger than levels of seed flow and pollen and seed gene flow vary independently across species (ibid.). Moreover, Petit *et al.* (2005) show that measures of subdivision that take into account the degree of similarity between haplotypes (N_{ST}, or R_{ST}) make better use of the information inherent in haplotype data than standard measures based on allele frequencies only.

Neutral organelle markers can be extremely useful for phylogeographic studies (Petit and Vendramin, 2005). The phylogeographic structure of forest tree species can be influenced by several factors, among which history during the glaciations and in the post-glacial period, life history traits of the species and human impact are assumed to have played a major role. Phylogeography can provide essential background information to disentangle current from past processes and to understand the consequences of crucial events such as colonization in the life and longevity of plant species (Petit *et al.*, 2003). The comprehension of the past dynamics of diversity can be extremely useful to predict the possible future migrations related to the expected climate changes (Pitelka *et al.*, 1997). In conservation and management of genetic resources, phylogeographic studies may help identify key regions deserving priority for conservation (Petit *et al.*, 2003). A phylogeographic survey may allow tracing of wood and other plant products, providing tools to combat illegal logging or to label products originating from sustainably managed regions (Deguilloux, Pemonge and Petit, 2002). Finally, the background on seed flow to emerge from phylogeographic surveys can be used to evaluate risks associated with the use of transplastomic plants (Petit and Vendramin, 2005; Daniell *et al.*, 1998).

Technology is also rapidly evolving in neutral marker analysis, moving from markers such as microsatellites towards single nucleotide polymorphism (SNP) markers due to cost, efficiency and automation considerations. Because of the different characteristics of the two markers systems in terms of mutation processes and rates, they will both find use in ecological studies.

14.3 REFERENCES

Daniell, H., Datta, R., Varma, S., Gray, S. & Lee, S. 1998. Containment of herbicide resistance through genetic engineering of the chloroplast genome. *Nature Biotechnology*, 16: 345-348.

Deguilloux M-F., Pemonge M-H. & Petit, R.J. 2002. Novel perspectives in wood certification and forensics: dry wood as a source of DNA. *Proceedings of the Royal Society of London*, B, 269: 1039-1046.

Petit, R.J., Aguinagalde, I., de Beaulieu, J.L., Bittkau, C., Brewer, S., Chedaddi, R., Ennos, R., Fineschi, S., Grivet, D., Lascoux, M., Mohanty, A., Muller-Starck, G., Demesure-Musch, B., Palme, A., Martin, J.P., Rendell, S. & Vendramin, G.G. 2003. Glacial refugia: hotspot but not melting pots of genetic diversity. *Science*, 300: 1563-1565.

Petit, R.J., Duminil, J., Fineschi, S., Hampe, A., Salvini, D. & Vendramin, G.G. 2005. Comparative organization of chloroplast, mitochondrial and nuclear diversity in plant populations. *Molecular Ecology*, 14. (available at www.blackwell-synergy.com/doi/abs/10.1111/j.1365-294X.2004.02410.x)

Petit, R.J. & Vendramin, G.G. 2005. Phylogeography of organelle DNA in plants: an introduction. *In* S. Weiss & N. Ferrand, eds. *PhyloGeography of southern European refugia*. Amsterdam, Kluwer. (in press)

Pitelka, L.F. & The Plant Migration Workshop Group. 1997. Plant migration and climate change. *American Scientist*, 85: 464-473.

IV. Debating the issues

15. Background document to the e-mail conference on the role of biotechnology for the characterization and conservation of crop, forest, animal and fishery genetic resources in developing countries

John Ruane and Andrea Sonnino

15.1 INTRODUCTION

On the occasion of World Food Day 2004, the United Nations Secretary-General Kofi Annan urged "individuals and institutions alike to give greater attention to biodiversity as a key theme in our efforts to fight the twin scourges of hunger and poverty and achieve the Millennium Development Goals." He also noted that the unprecedented loss of biodiversity over the past century was a major cause for alarm, where "many freshwater fish species, which can provide crucial dietary diversity to the poorest households, have become extinct, and many of the world's most important marine fisheries have been decimated. Food supplies have also been made more vulnerable by our reliance on a very small number of species: just 30 crop species dominate food production and 90 percent of our animal food supply comes from just 14 mammal and bird species - species which themselves rely on biodiversity for their productivity and survival. There has been a substantial reduction in crop genetic diversity in the field and many livestock breeds are threatened with extinction (www.un.org/News/Press/docs/2004/sgsm9539.doc.htm)".

On the same occasion, FAO's Director-General Jacques Diouf also underlined that although forests are among the world's most important repositories of biological diversity, the world forest cover is decreasing at an alarming rate (www.fao.org/wfd/2004/dgmessage_2004_en.asp).

It is in this context of declining agricultural biodiversity that the FAO Biotechnology Forum (www.fao.org/biotech/forum.asp) hosted an e-mail conference from 6 June to 3 July 2005 to consider the role that biotechnology can play in the characterization and conservation of crop, forest, animal and fishery genetic resources in developing countries. Biotechnology is a broad collection of tools that can be applied for a range of different purposes (e.g. genetic improvement of populations; disease diagnosis and vaccine development; and improvement of feeds). This conference focused on biotechnology tools such as molecular markers, cryopreservation and reproductive technologies that can be used directly for the characterization and/or conservation of genetic resources for food and agriculture. Genetic modification and GMOs will not be considered here.

On 5-7 March 2005, as part of the preparations for this e-mail conference, an international workshop was held in Turin, Italy, entitled "*The Role of Biotechnology for the Characterisation and Conservation of Crop, Forestry, Animal and Fishery Genetic Resources*", organized by the FAO Working Group on Biotechnology, the Fondazione per le Biotecnologie, the Econogene project and the Società Italiana di Genetica Agraria. Proceedings of the workshop (available at www.fao.org/biotech/torino05.htm) include 20 papers and 37 poster abstracts covering applications of molecular markers, cryopreservation and reproductive technologies and can be consulted by anyone looking for more detailed technical information in this area. In addition to the poster session, the workshop was split into Session I on the status of the world's agricultural biodiversity, Session II on the use of biotechnology for conservation of genetic resources, and Session IV on genetic characterization of populations and its use in conservation decision-making. Session III presented results from Econogene, a European Union-funded project combining a molecular analysis of biodiversity, socio-economics and geostatistics to address the conservation of sheep and goat genetic resources and rural development in marginal agrosystems in Europe.

This chapter provides background information for the e-mail conference. First, a brief overview of genetic resources for food and agriculture is provided (Section 15.2), followed by more specific information regarding the current status of the genetic resources in the different food and agricultural sectors (Section 15.3). A brief description of the relevant biotechnologies is then given (Section 15.4), followed by a discussion of some key issues and some questions that might be addressed in the e-mail conference (Section 15.5).

15.2 A BRIEF BACKGROUND TO GENETIC RESOURCES FOR FOOD AND AGRICULTURE

As described in Diamond (2002), domestication of crops and livestock, where species selected from the wild adapt to a special habitat created for them by humans, was a momentous development with major consequences for human societies; since "it provides most of our food today, it was prerequisite to the rise of civilization, and it transformed global demography."

Domestication seems to have arisen independently on at least five different occasions in different parts of the world where, for example, hunter-gatherers in the Fertile Crescent (an area in Asia, including parts of Iran, Iraq, Israel, Jordan, Lebanon, Syria and Turkey) began domesticating barley, peas, wheat, cows, goats, pigs and sheep about 10 000 years ago. Occasionally the same species (e.g. cow, sheep) was domesticated independently in different places. The process of domestication has continued ever since, for example, in the Middle Ages with rabbit and strawberry.

Domestication had major impacts on human societies, to enable the transition from a hunter-gatherer society to a settled farming society. It also led to major changes in the species that were domesticated. Since their domestication, crops and livestock have been used in almost all environments of the world and many have been domesticated in particular environments for very long periods of time. Soay sheep on Hirta, off the west coast of Scotland, for example, are thought to be the survivors of the sheep first brought to Britain by the first farmers 5 000 years ago. Each environment has a certain set of stressors that impacts the production environment. If we consider livestock, these might include climatic features (heat/cold, amount of rain, humidity), feeding conditions (quality/quantity vegetation available, access to water) or diseases (parasitic infections, etc.), and over time, populations adapted to the environment and its specific stressors. Farmers have also exercised a major influence on the domesticated species by preferentially selecting for specific morphological or production traits.

Unlike crops and livestock, domestication is generally a very recent phenomenon in the fishery and forestry sectors. Only a small number of forest tree species have been domesticated to some degree, and only over the past half century. Such tree species, used in plantation forestry, are usually no more than one to two generations removed from their "wild relatives" (see, for example, Sigaud, 2005). Concerning fish species, with the exception of a few species such as common carp domesticated about 2 000 years ago, aquaculture (fish farming) is a relatively new development (Bartley, 2005) and most of the fish consumed today still come from wild populations, although it is predicted that by 2030 consumption from the ever-growing aquaculture industry will exceed capture fisheries.

In order to consider the agricultural genetic resources in this e-mail conference, a wide range of species has therefore been included, from those domesticated 10 000 years ago to those domesticated in recent times, as well as those that have not been domesticated but that are nevertheless of importance for mankind's food and agriculture.

As will be described in more detail in Section 15.3, many of these genetic resources are endangered. The need to conserve them is now widely accepted and generally justified for one or more of a variety of reasons: their importance in ensuring against future changes in market needs and production conditions; as a source of material for scientific research and future germplasm development; and as a cultural and historical part of our heritage, passed down from previous generations (e.g. Ruane, 2000). The issue has received considerable attention at the international policy-making level. For example, it is addressed under the Convention on Biological Diversity (CBD), a legally-binding agreement with 188 parties, through its agricultural biodiversity work programme (www.biodiv.org/programmes/areas/agro/default.asp), and under the International Treaty on Plant Genetic Resources for

Food and Agriculture, a legally-binding treaty with 69 parties that entered into force in June 2004 (www.fao.org/ag/cgrfa/itpgr.htm).

Two major strategies for conservation can be distinguished: *in situ* conservation where a population is maintained in its natural or agricultural habitat and *ex situ* conservation, where it is maintained outside of this habitat. *Ex situ* conservation can involve living individuals (e.g. animals in zoos, fish in aquaria, crops or trees in botanical gardens and arboreta), or tissues/genetic material from individuals (such as seeds, pollen, sperm, embryos and DNA). *Ex situ* conservation is important for crops, where an estimated 6 million plant, mainly crop, accessions (i.e. samples of a variety collected at a specific location and time) are stored in national, regional and international collections (FAO, 1997, Chapter 3). Of the 6 million accessions, over 80 percent are in national collections and around 90 percent are stored in seed gene banks which, depending on the seed storage facilities, may allow short-, medium- or long-term storage. A gene bank is a storage facility where germplasm of plant or animal origin is stored in forms such as seeds, pollen, semen or embryos; in *in vitro* culture (germplasm kept as sterile plant tissues or plantlets on nutrient gels); in cryogenic storage; or in a field gene bank, i.e. as plants growing in the field. *In situ* conservation is currently more important than *ex situ* conservation for the livestock, forestry and fisheries sectors.

Characterization of genetic resources goes hand in hand with their conservation since it is fundamental both for understanding what is being conserved and for choosing the genetic resources that should be conserved. Characterization enables us to identify the key features, both the strengths and weaknesses, of the available genetic resources and this knowledge can also be used to develop breeding programmes for sustainable use of the genetic resource to harness and/or disseminate the positive attributes identified in the population. Characterization can also play an important role regarding issues of access to and benefit-sharing of agricultural genetic resources, as well the promotion and control of "bioprospecting" (i.e. biodiversity prospecting, the search for commercially valuable compounds, substances or genetic material in nature). If a country has characterized its genetic resources, it should be appropriately positioned to develop and implement conservation strategies for targeted species. It should also be able to promote and oversee any potential commercial or other enterprises arising from the knowledge generated to protect the resources (or any of their products) from inappropriate bioprospecting by foreign companies or countries. It should also be able to negotiate appropriate compensation for use of these resources by third parties, be they national or foreign.

The kinds of features that could be characterized for the different agricultural populations include their morphological and physiological features; their phenotypic and economic performance in different environments; their interaction with and relation to the environment; their value as social or cultural objects; their role in scientific research; their diversity at the gene and DNA sequence level; their population size and their degree of endangerment.

15.3 STATUS OF GENETIC RESOURCES IN THE DIFFERENT FOOD AND AGRICULTURAL SECTORS

This section aims at providing an overview of the current situation regarding genetic resources in each of the four sectors.

15.3.1 Crop genetic resources

Compared to the large numbers of wild plant species available, only very few were actually domesticated (Diamond, 2002). Although this number is small, just a subset of them, again, are of special importance globally. For example, considering plant-derived energy intake worldwide, just three crops (wheat, rice and maize) provide over 50 percent of the intake and 30 crops provide 95 percent (FAO, 1997, Chapter 1). Splitting the world into 17 subregions, there are only 12 major crops that supply more than 5 percent of the plant-derived energy intake in one or more subregions, i.e. wheat, rice, maize, millet, sorghum, potato, sugar cane, soybean, sweet potato, cassava, the common bean and related species (*Phaseolus*), and banana/plantain. It should be kept in mind, however, that although some specific crops may not have any global/regional significance, they may nevertheless be important at the individual country level (e.g. tef in Ethiopia).

Although the number of major crop species is quite small, as seen above, the diversity within each species can be very substantial. Within a species, cultivated crop varieties (cultivars) can be broadly categorized as farmer's varieties (also known as landraces or traditional varieties, i.e. populations that are the product of breeding or selection carried out by farmers, either deliberately or not, continuously over many generations) and modern varieties (also known as high-yielding or high-response varieties, which are the products from professional breeders working in publicly-funded research institutes or private companies). Farmer's varieties tend to maintain high levels of genetic diversity and to be the focus of conservation efforts (FAO, 1997).

The first comprehensive worldwide report on the status and use of crop genetic resources was published by FAO in 1997 (www.fao.org/ag/AGP/AGPS/Pgrfa/pdf/swrfull.pdf). This *State of the World's Plant Genetic Resources for Food and Agriculture (SoW-PGRFA)* was developed through a participatory country-driven process and the primary data sources for the publication were detailed country reports from a total of 154 countries. Annex 2 of the report (FAO, 1997) documents the state of diversity within each of the 12 major crops. It shows, for example, that the global *ex situ* collection of millet contains roughly 90 000 accessions, of which 2 percent are wild relatives, 33 percent landraces and old cultivars, and 5 percent advanced cultivars and breeding lines, while 60 percent are unknown and mixed. Although incomplete data makes it difficult to give a definitive picture of the kind of material stored in the world's *ex situ* collections, it is estimated that roughly half the accessions for which the type of material is known are advanced cultivars and breeding lines, while a third are landraces and around 10 percent are wild species (FAO, 1997, Chapter 3).

In their country reports nearly all countries stated that genetic erosion (i.e. the loss of genetic diversity, generally as a result of social, economic and agricultural changes) was a serious problem. The main cause cited was the replacement of local varieties or landraces by improved and/or exotic varieties and species, and numerous examples were provided (e.g. displacement of traditional barley and durum wheat varieties by introduced varieties in Ethiopia). Other important causes cited were: overexploitation of plant genetic resources, including overgrazing and reduced fallow periods in shifting cultivation; deforestation and land clearance; population pressure and urbanization; environmental degradation (e.g. desertification, flooding); and changes in agricultural systems (FAO, 1997, Chapter 1). Preparations are underway for the second *Report on the SoW-PGRFA*, which should be finalized in 2008.

15.3.2 Forest genetic resources

Most forest tree species are characterized by inherently high levels of variation and extensive natural ranges. They are long-lived, frequently outcrossing organisms with generally large distribution ranges. A high level of intra-specific genetic variation is needed to ensure present-day and future adaptability to changing environmental conditions. It is also needed to maintain options and potential for improvement to meet changing end-use requirements. Forests provide a wide range of goods and services such as timber, fibre, fuelwood, food, fodder, gum, resins, medicines, pharmaceutical products and environmental stabilization. Similar goods and services are often provided by a wide range of genera and tree species. Despite the availability of a large number of forest tree species, less than 500 have been systematically tested for their present-day utility for human beings and less than 40 species are included in intensive selection and breeding programmes (FAO, 2000a).

Estimates for the year 2000 indicated that the world's forest cover was about 3.9 billion ha, of which 47 percent was in the tropics, 33 percent in the boreal zone (i.e. the northernmost forest zone), 11 percent in temperate areas and 9 percent in the subtropics (FAO, 2001). About 95 percent of the forest cover was in natural forest and 5 percent in forest plantations (defined as forest stands established by planting or/and seeding in the process of afforestation or reforestation). Of the roughly 190 million ha of forest plantation, 48 percent had an industrial end-use (e.g. pulpwood for paper), 26 percent a non-industrial purpose (e.g. fuelwood, soil and water conservation), and for 26 percent it was not specified.

Most of the forest genetic resources worldwide are therefore found in natural, largely unmanaged forests which, however, are frequently intensively used for the provision of both wood and non-wood products for local, national and international use. Conversion of forests to other land uses represents the greatest threat to forests and their diversity. According to FAO (2001), "increasing pressure from human populations and aspirations for higher standards of living, without due concern to the sustainability of the resources underpinning such developments, heighten these concerns. While some land use changes are inevitable, it is important that such changes be planned and managed to address complementary goals. Concerns for biological and genetic conservation should be major components of land use planning and forest management strategies". Unlike the situation in crops, however, no global overview of the status of forest tree genetic resources has yet been carried out, although the idea of developing a first global assessment of forest genetic resources is currently being discussed (Sigaud, 2005).

15.3.3 Animal genetic resources

Of the 50 000 known bird and mammalian species, around 30 have been used extensively for agricultural purposes, with fewer than 14 accounting for over 90 percent of global livestock production (FAO, 1999). In the millennia following their domestication, animal species were influenced by both natural and artificial selection in a wide range of geographic and climatic areas and production systems, leading to the development of large numbers of genetically diverse domestic animal breeds.

Many of these breeds are currently endangered. In developing countries, the introduction of exotic breeds and their spread through indiscriminate crossbreeding is a major reason for loss of local, indigenous breeds. Other factors include changes in breeders' preferences due to short-term socio-economic influences, degradation of the ecosystem in which the breeds were developed, natural disasters (e.g. drought and diseases), wars and other forms of political unrest and instability (FAO, 1998).

In 1993, FAO published the first *World Watch List for Domestic Animal Diversity*, which provides a basic global overview of the state of the genetic resources of seven mammalian species (buffalo, cattle, donkey, goat, horse, pig and sheep). The third edition, expanded to cover 30 mammalian and bird species, showed that of the over 4 000 breeds with population data, roughly 30 percent could be classified at a high risk of loss and that chicken and cattle have large numbers of breeds at risk while horses and geese have the highest percentages of breeds at risk of loss (FAO, 2000b). In addition, with the exception of the wild boar and wild red jungle fowl (ancestors of the domestic pig and chicken respectively), the putative wild ancestors of the major livestock species are in danger of extinction or are already extinct (Hanotte and Jianlin, 2005).

In a similar process to that described for the crop sector, FAO is preparing the first *State of the World's Animal Genetic Resources*, which is expected to be completed by 2006 and for which 151 countries have accepted to submit country reports to date. The aim of the country and global assessments is to provide a comprehensive analysis of the status and trends of the world's farm animal biodiversity and of their underlying causes, as well as of local knowledge regarding its management (Cardellino, 2005).

15.3.4 Fishery genetic resources

Like forestry but unlike crops and livestock, wild populations currently play a greater role than domesticated populations in the fishery sector. When considering the status of the world's fishery genetic resources, both the aquatic species that are fished (capture fishing) and those that are farmed (aquaculture) can be included. Bartley (2005) gives an overview of the status of aquatic species that are fished or farmed worldwide.

For capture fisheries in 2002, FAO Members reported that 974 taxa (i.e. taxonomic units such as a family, genus or species) of fin-fish, 143 taxa of crustacea, 114 taxa of molluscs, 26 taxa of plants and 73 taxa of miscellaneous animals (such as sea urchins, sea cucumbers and marine mammals) were taken from the world's capture fisheries. These numbers are likely to be underestimates, however, because reporting is often incomplete. Although over 1 000 taxa are represented, about ten species make up about one-third of capture fisheries production. Unlike the crop and animal sectors, there is no systematic effort to describe the state of the world's fishery genetic resources below the species level. Nevertheless, description of the fish genetic stocks at the within-species level is important for fisheries management and is an active research area (Bartley, 2005; Primmer, 2005).

For aquaculture in 2002, FAO Members reported that apart from a small number of aquatic plants and animals, a total of 153 species of fish, 60 species of molluscs and 44 species of crustaceans were farmed. The relative novelty of aquaculture can be shown by the fact that its contribution to global supplies of fish, crustaceans and molluscs has grown from just 4 percent in 1970 to 30 percent in 2002 (Bartley, 2005).

Aquatic science in general lags behind terrestrial sciences in identifying species, understanding ecosystem relationships and assessing potential uses for genetic resources. Communities of life on the ocean floor are poorly understood, and every year some aquatic species become extinct before they have even been identified. Fish genetic resources worldwide are currently threatened by overfishing, but also by habitat destruction (e.g. dam building) and pollution from human activities (Greer and Harvey, 2004).

Every two years FAO publishes the *State of World Fisheries and Aquaculture*, which, *inter alia*, provides a global overview of capture fisheries (marine and inland) and aquaculture. Regarding exploitation of marine fish stocks (estimated production at 84.5 million tonnes in 2002), the most recent edition indicates that there has been a consistent downward trend since 1974 in the proportion of stocks offering potential for expansion, together with an increase in the proportion of overexploited and depleted stocks, rising from about 10 percent in the mid-1970s to close to 25 percent in the early 2000s. The percentage of stocks exploited at or beyond their maximum sustainable levels varies widely among fishing regions. The report says that the information available continues to confirm that despite local differences, the global potential for marine capture fisheries has been reached, and more rigorous plans are needed to rebuild depleted stocks and prevent the decline of those being exploited at or close to their maximum potential (FAO, 2004a).

Inland fish stocks (estimated production was 8.7 million tonnes in 2002), unlike the major marine fish stocks, are less well defined and occur over much smaller geographical areas such as individual lakes, rice fields or rivers, or over vast areas such s transboundary watersheds that are often situated in areas that are difficult to ccess. These factors make it costly to monitor the exploitation and status of fish tocks and, in fact, very few countries can afford to do so. As a result, most countries eport only a small fraction of their catch of inland fisheries by species, further ompounding the problem of accurate assessment. There are indications, however, hat these resources are undervalued and threatened by habitat alteration, egradation and unsustainable fishing activities (FAO, 2004a).

5.4 OVERVIEW OF RELEVANT BIOTECHNOLOGIES

As seen in Section 15.3, a wide range of genetic resources are currently being used by umankind for food and agricultural purposes and many of them are in danger of eing lost for reasons such as over-exploitation (fish), replacement of local with nternational germplasm (crops/livestock), habitat change and destruction crop/fish/forest/livestock). A brief overview will be presented of molecular markers nd cryopreservation, *in vitro* culture and reproductive technologies and how they an be used for the characterization and/or conservation of these genetic resources.

5.4.1 Molecular markers

'he importance of characterizing genetic resources was described in Section 15.2 as vell as the wide range of features that can be characterized in each population, such as norphology, phenotypic performance and degree of endangerment. Whereas henotypes (e.g. yield, growth rate) or morphological traits (coat colour, seed shape) re influenced by both genetic and environmental factors, the use of molecular narkers reveals differences at the DNA level that are not influenced by the nvironment (de Vicente *et al.*, 2005). Molecular markers usually do not have any iological effect and are normally assumed to represent neutral loci unaffected by election. They are identifiable DNA sequences found at specific chromosomal ocations on the genome and transmitted by the standard laws of inheritance from one eneration to the next. A number of types of molecular markers exist, such as RFLP narkers, random amplified polymorphic DNA (RAPD) markers, amplified fragment ength polymorphism (AFLP) markers, and microsatellites and single nucleotide olymorphism (SNP) markers, which can differ in a variety of ways, such as the mount of time, money and labour needed or the amount of genetic variation found at ach marker in a given population. Microsatellites, for example, are simple DNA equences, usually two or three DNA bases long, repeated a variable number of times n tandem. (For more background information on molecular markers, see FAO, 2003).

Molecular markers are used in a variety of approaches to characterize and onserve genetic resources, the most important of which will be described below. ome of the approaches are applied in each of the crop, forestry, animal and fishery ectors (e.g. estimating the genetic relationships between populations within a pecies), while others are more sector-specific (e.g. identifying duplicate accessions in rop gene banks or monitoring effective population sizes in capture fish opulations).

Estimating the genetic relationships between populations within a species

This is one of the major uses of molecular markers in agricultural research Molecular markers are typed in populations of the same species (for instance, thos located in different places) and about which inferences can be made on hov genetically related they are to each other by comparing the frequencies anc presence/absence of marker loci in the different populations. The more difference in markers found between the populations, the greater the genetic distance betwee them, and the populations are inferred to be less genetically related to each othe than to populations that have more similar marker loci frequencies. Taking a simpl example for illustrative purposes (in practice many markers would be use simultaneously), if individuals in populations A, B and C are typed for a molecula marker M (with two alternative forms or alleles called M1 and M2) and a individuals typed in populations A and B have allele M1 while those in populatio C have M2, this suggests that A and B are genetically more similar to each othe than to population C. In using molecular markers to estimate genetic distance between populations, it is assumed that they represent neutral loci (i.e. do nc reflect genetic differences between populations at genes that are or have been unde natural or artificial [farmer] selection) and that the use of a relatively small numbe of independently segregating marker loci will be a good predictor of the overa genomic diversity of a population (Hanotte and Jianlin, 2005).

Typically, a genetic diversity study involves sampling biological tissue (e.g. bloo leaves, fish scales) from a certain number of individuals from a number c populations of interest and typing them in the laboratory for a number of molecul marker loci. A statistical analysis of the marker data is then carried out to estima the genetic relationships between the populations. For example, Lanteri an Barcaccia (2005) sampled 30 individuals each from seven populations of wi cardoon identified in Sardinia and Sicily, and each individual was genotyped fc 32 microsatellite markers as well as seven AFLP primer combinations. They used statistical analysis method called principal coordinate analysis to show that th Sardinian and Sicilian populations were clearly differentiated, representing tw distinct gene pools.

Such studies have been routinely applied in each of the agricultural sectors. Fo example, in developing a management strategy for commercially exploited fis molecular markers are used to characterize the within-species genetic structure of t populations being harvested to determine the units between which limited gene flo occurs, i.e. to identify units that should not be overfished (Primmer, 2005). A rece questionnaire-based survey of animal diversity studies carried out during the last t years provided information on a total of 86 projects involving 13 livestock speci from 93 countries (FAO, 2004b). Most of the projects were for ruminants, especial cattle. Blood was the most frequently used biological sample and microsatellites we the most commonly used molecular markers.

As pointed out by Lenstra *et al.* (2005), these kinds of studies are not on relevant for the purposes of characterization of genetic resources, but they can al provide useful insights into the history of domestication. Evidence from molecular markers has been used often to elucidate whether individual crop or livestock species were domesticated just once at a single site or independently several times in different parts of the world (Diamond, 2002).

Establishing and managing gene banks

As described in Section 15.2, gene banks are an important tool for conservation of crop genetic resources, although currently of more limited importance for livestock, forestry and fish. Molecular marker information can be used in a number of gene bank-related activities, such as sampling of material, management of the collections and stimulating use of germplasm stored therein (Lanteri and Barcaccia, 2005).

When sampling germplasm to create a gene bank, information from the use of molecular markers to assess the within- and between-population genetic diversity (as described in the section above) can be used in conjunction with data on other characteristics of the populations to assist decision-making regarding which material should be sampled (e.g. Simianer, 2005). Since the amount of germplasm to be stored may be limited, for example, due to financial reasons, it would be important to sample material from genetically diverse individuals.

Once a gene bank has been established, molecular markers can also be used to assist in a number of gene bank management activities. As a first example, regeneration of seeds or other reproductive plant material in storage is an important task in a gene bank, and the regeneration requirements (how frequently accessions need to be regenerated) depend on factors such as the species concerned, the storage conditions and the quality of the individual accessions (FAO, 1997, Chapter 3). Markers can be used to check whether changes in alleles or in allele frequencies are taking place over time as a result of the adopted conservation strategies, for instance of frequent regenerations (de Vicente *et al.*, 2005). Second, while safety duplication of unique accessions acts as an insurance against possible accidents, unintentional duplication or overduplication of accessions is wasteful, however (FAO, 1997), and markers can be used to detect them (de Vicente *et al.*, 2005). Third, to reduce costs and/or encourage greater and more efficient use by farmers and breeders, core collections (i.e. a subset of accessions selected to contain the maximum available variation in a small number of accessions) have been established for many crop species. Again, molecular markers can assist in this task.

Finally, it is argued (e.g. Ferreira, 2005) that molecular markers offer opportunities to increase the use of crop genetic resources from gene banks and will therefore stimulate increased use of gene bank material for research purposes, thereby playing an important role in the sustainable conservation of these genetic resources. Since a major justification for the existence of germplasm collections is use of the conserved accessions, de Vicente *et al.* (2005) underline the importance of genetically characterizing them so that valuable genes can be identified for use in developing new plant varieties by conventional and advanced breeding methodologies.

Gene flow from domesticated populations to wild relatives

Wild relatives of domesticated species are of conservation interest because they represent one of the components of agricultural biodiversity. They have also proven useful for genetic improvement of a range of cultivated varieties (FAO, 1997, Chapter 1). As described in FAO (2002), crossing of domesticated populations with their wild relatives has been well documented in some crop, forest tree, animal and fish species. For example, in crops, gene flow has been observed between rice and perennial rice, between maize and teosinte, and between sugar beet and wild beet, while in animals there is evidence for crossing of domestic cattle with wild North American bison, and of domestic pigs with European wild boars. In forestry, gene flow from intensively bred forest trees to natural populations of the same or closely related species ("wild relatives") has been a cause for concern in many species, for example, the European black poplar (*Populus nigra*). For fish, crossing of escaped farmed Atlantic salmon with wild Atlantic salmon is also a much-discussed problem.

Gene flow from a domesticated population will cause changes in the genetic diversity of wild populations. In cases where diversity is systematically reduced over time by such gene flow, it may potentially lead to genetic extinction of the wild population in its original genetic state (see, for example, Papa, 2005). Molecular markers can assist in conservation of the wild relatives through their use to distinguish hybrids from non-hybrid wild relatives and to monitor gene flow from domesticated populations to their wild relatives.

Estimating and monitoring the effective population size

When developing conservation strategies for wild species, accurate estimates of the effective population size (Ne) are important for predicting a number of parameters such as the rate of inbreeding (Primmer, 2005). Knowing the rate of inbreeding is vita because increases in inbreeding (due to mating of genetically related individuals) ca lead to the reduced abilities of individuals to survive and reproduce (a phenomeno called inbreeding depression). Ne can be estimated using pedigree information and/o knowledge about the population breeding structure (male/female mating ratio, famil sizes, etc.). However, for wild populations, especially of aquatic species, it can b difficult to get this information. Molecular markers can be used to estimate Ne, whic has been done, for example, in several commercial fish populations (Primmer, 2005). Similarly, molecular markers can also be used as a tool to monitor fish population size over time and to detect any major recent declines in Ne.

15.4.2 Cryopreservation, *in vitro* culture and reproductive technologies

As described in Section 15.2, there are two major strategies for conservation, *in sit* and *ex situ*, the latter involving, for instance, storage of genetic material in gene bank keeping live animals in zoos, or keeping plants in botanical gardens. Biotechnolog through cryopreservation (freezing) technologies and *in vitro* culture, provid additional tools for *ex situ* conservation.

Cryopreservation involves the preservation of germplasm in a dormant state by storage at ultra-low temperatures, usually in liquid nitrogen (-196 °C), and can be used to preserve biological material (e.g. seeds, sperm, embryos) of crop, livestock, forest or fish populations indefinitely in gene banks. It allows large quantities of genetic material to be stored in a limited amount of space. If the living populations later become extinct, they can in many instances be regenerated using the frozen genetic material. In addition, if the living populations are declining in numbers and develop problems due to inbreeding, these problems can be alleviated by introducing unrelated genetic material from the gene bank. In animals, collection of germplasm for cryopreservation and production of live offspring from cryopreserved genetic material depends on a number of associated reproductive technologies, such as embryo transfer. *In vitro* culture is a plant biotechnology that involves application of slow growth procedures to germplasm accessions kept as sterile plant tissues or plantlets on nutrient gels.

Here below, we will provide a brief overview of the current status of these biotechnologies, focusing on cryopreservation since it is applicable in all four sectors. Information provided on cryopreservation in livestock/fish and crops/forest trees is based primarily on Hiemstra *et al.* (2005) and Panis and Lambardi (2005), and the references therein.

Livestock and fish

Before considering the main types of genetic material that have been cryopreserved, a brief overview of relevant terminology may be useful. In sexual reproduction, through fertilization there is a fusion of two haploid gametes to form the diploid zygote, a cell that begins dividing and develops to become an embryo. (Note: haploid and diploid cells have one or two sets of chromosomes, respectively.) The female gamete is called the ovum or egg, and is derived from a cell called an oocyte that has undergone two meiotic divisions. The male gamete is called the sperm, contained in fluid called semen.

For cryopreservation purposes, semen is currently the biological material of greatest importance, although embryos and, to a lesser degree, oocytes and somatic cells, have also received considerable research attention. Success in freezing these materials varies considerably from species to species, which is not surprising given the wide diversity in reproductive systems (e.g. fertilization is usually external in fish and internal in birds where the female stores the sperm before fertilization). Collection of biological materials for the gene bank and production of live offspring following thawing of formerly cryopreserved material is made possible through a variety of reproductive technologies, such as artificial insemination (AI) using semen, multiple ovulation and embryo transfer using embryos.

Freezing of semen and subsequent successful AI with thawed semen was first achieved over 50 years ago and semen of most livestock species can now be frozen adequately. There are large differences between species in insemination techniques and pregnancy rates using fresh or frozen semen, where AI with frozen semen is most successful in cattle. For domesticated birds, although AI is less widely used than in mammals, reasonable success rates with frozen semen, albeit at times highly variable, have been reported for the major bird species. For fish, sperm cryopreservation has been tested in over 200 fish species with external fertilization and the present state of the art for many species of fish seems to be adequate for the purpose of gene banking (Hiemstra *et al.*, 2005).

For embryos, cryopreservation in cattle is now a routine procedure. It has also been reported for other mammalian species, such as horse, pig, rabbit and sheep (ERFP, 2003). In order for embryos to be cryopreserved, they can be collected using non-surgical (cattle and horses) or surgical methods (other mammalian species). Inducing multiple ovulation using hormones can increase the efficiency of the process. Alternatively, embryos can be collected through *in vitro* maturation and *in vitro* fertilization of oocytes, a more technically demanding approach. Production of live offspring following cryopreservation involves non-surgical (only in cattle and horses and, to a lesser extent, pigs) or surgical embryo transfer to recipient animals. For bird and fish species, however, cryopreservation of embryos has not been successful, mainly because of the large size, the high lipid content and the polar organization of the ova and early embryos.

For mammalian oocytes, live offspring have been reported in cattle and horse from embryos produced using cryopreserved oocytes, although the efficiency and reliability of using oocytes for generating offspring is still much lower today than with cryopreserved embryos. Oocytes for cryopreservation in the gene bank can be collected from ovaries of dead animals or by ovum pick-up (recovery of oocytes using an ultrasound probe) on live animals. Production of live offspring following thawing of cryopreserved oocytes involves *in vitro* maturation and fertilization, and eventual embryo transfer to recipient animals. Cryopreservation of bird and fish oocytes has not yet been successful, mainly for the same reasons given earlier for embryos.

For somatic cells (i.e. cells not involved in sexual reproduction), a number of cell types (e.g. skin fibroblasts, mammary cells) can now be cryopreserved. Collection of material for the gene bank is easy and cheap. Production of live offspring (which, with the exception of mitochondrial DNA, are genetic clones of the individual that donated the somatic cells) is complicated, however, and has a low rate of success. It involves culturing the thawed cells and then transferring their nuclei (i.e. the organelle that contains the genetic material) to, or fusion of the somatic cells with enucleated oocytes (i.e. without a nucleus) from another individual. The resulting embryos are then cultured and eventually transferred to recipient animals. In mammals, live offspring have been obtained from embryos generated from somatic cells in a number of mammalian species (cattle, goats, horse, pigs, rabbits and sheep) and in fish (zebrafish), but not yet in poultry. The low success rates, as well as the many developmental problems encountered, are current major limitations to application of the technology.

Crops and forest trees

As described in Section 15.2, about 90 percent of the six million plant accessions in gene banks, mainly crops, are stored in seed gene banks. However, storage of seeds is not an option for crops or trees that do not produce seed, such as bananas, or that produce recalcitrant or non-orthodox seed (i.e. seed that does not survive under cold storage and/or the drying conditions used in conventional *ex situ* conservation), such as mango, coffee, oak and several tropical forest tree species. The same applies to plant species that are propagated vegetatively to preserve the unique genomic constitution of cultivars, such as fruit and several timber and ornamental trees. It also presents practical problems in most forest tree species in regard to the necessary periodic regeneration of seedlots, because tree species usually have a long vegetative period prior to producing flowers and seed (lasting from several years to several decades), and because they require large areas to carry out such an operation due to outcrossing and their size.

Genetic resources of these species are normally conserved either *in vivo*, i.e. living collections, or by *in vitro* culture, where slow growth procedures allow the plant material to be held for 1-15 years under tissue culture conditions with periodic regeneration (subculturing), depending on the species. Normally, growth is limited using low temperatures often in combination with low light intensity or even darkness. Temperatures in the range of 0-5 °C are employed for cold-tolerant species and 15-20 °C for tropical species. Growth can also be limited by modifying the culture medium and reducing oxygen levels available to the cultures (Rao, 2004). The former option (*in vivo*) is the cheapest and thus the most widely adopted, but exposes the material to the harms of the external environment, e.g. contamination with pathogens or parasites. *In vitro* culture, on the other hand, ensures that the collections are kept free from pest and diseases and protected from adverse conditions. Maintenance of *in vitro* collections is labour-intensive, however, and there is still a risk of losing accessions due to contamination, human error or somaclonal variation, i.e. mutations that occur spontaneously in tissue culture, with a frequency that increases with repeated subculturing (Panis and Lambardi, 2005).

In all these situations, as well as for long-term storage of seeds from orthodox species, cryopreservation offers an alternative tool or complementary strategy for *ex situ* conservation. Following plant cell, tissue or organ storage at low temperatures, plants can then be regenerated and there seems to be no clear evidence so far of morphological, cytological or genetic alterations arising as a result of cryopreservation (Panis and Lambardi, 2005). They conclude that, however, despite the fact that cryogenic procedures are now being developed for an increasing number of recalcitrant seeds and a wide range of tissues and organs, the routine utilization of cryopreservation for the preservation of plant biodiversity is still limited. They note that the main drawback for a wider application of plant cryopreservation is the unavailability of efficient cryopreservation protocols for many plant species.

Panis and Lambardi (2005) nevertheless report that for various herbaceous (i.e. non-woody plants), hardwood (i.e. broadleaf, deciduous trees) and softwood species (i.e. coniferous trees), cryopreservation has been achieved of a wide range of tissues and organs such as cell suspensions, embryogenic cultures, pollen, meristematic tissues and seeds. Cell suspensions (cells in culture in moving or shaking liquid medium) have been cryopreserved from banana and grape while cryopreservation of embryogenic cultures (i.e. callus or suspension cultures with potential to differentiate somatic embryos) is an advanced technology in conifers and is already being successfully applied to numerous commercial species, including pines. There are methods for the cryopreservation of pollen from many crops, but its application is still rather limited and restricted to a few research centres. Meristematic tissues (also called meristems, i.e. plant tissues such as shoot tips, in which the cells are capable of active division and differentiation into specialized tissues such as shoots and roots) are the most common explants (i.e. a portion of a plant aseptically excised and prepared for culture in a nutrient medium) for the cryopreservation of vegetatively propagated species, such as fruit trees and many root and tuber crops. There is large-scale application of shoot tip cryopreservation in fruit crop germplasm collections, such as in plum and apple, while the number of cryopreserved herbaceous meristem samples is significantly smaller, although continuously growing. Seeds of most common agricultural and horticultural species can be cryopreserved, and for orthodox seeds this can be considered as an alternative since it offers long-term storage, to traditional storage (e.g. in celery).

15.5 SOME ISSUES AND QUESTIONS RELEVANT TO THE DEBATE

As with each conference hosted in the FAO Biotechnology Forum, the focus is on application of biotechnology in developing countries. In the debate on the role of biotechnology for the characterization/conservation of genetic resources for food and agriculture in developing countries, some of the potential factors that should be considered are briefly described below as they may influence applications of biotechnology for these purposes. In addition, a selection of some of the specific questions that participants might wish to address in the e-mail conference are also given.

15.5.1 Capacity issues

Some of the biotechnologies described in Section 15.4 require a considerable amount of infrastructure and capacity. For example, application of molecular markers to marine capture fishery populations requires appropriate sampling of individuals from the populations of interest, accurate laboratory typing of the samples and good statistical analysis of the data, *inter alia*. Some technologies that are and have been routinely applied in developed countries are rarely used in many developing countries. For example, AI is a commonplace tool and is used to breed most of dairy cows in the Nordic countries, but is little used in many developing countries.

A rough idea of the status of developing countries' capacities in some of the biotechnologies discussed in this chapter can be derived from a first analysis of about 2 000 crop sector entries from 71 countries in FAO-BioDeC (www.fao.org/biotech/

inventory_admin/dep/default.asp), a database providing information on biotechnology products/techniques in use or in the pipeline in developing countries (FAO, 2005). It reports over 400 molecular marker activities, most at the research level, in 43 developing countries. Considering the different regions, most activities were reported in Latin America, with 93 trials and 165 molecular marker projects at the research phase in nine countries. Species reported to be included in molecular marker programmes in the Latin American countries are sugar cane, rice, cocoa, banana, bean, maize and Andean local roots and tubers. The survey also indicates that most countries in Asia are undertaking a wide spectrum of crop research using molecular markers. Molecular marker-related research activities in Africa are reported to be underway in only a few of the countries, such as Ethiopia, Nigeria, South Africa and Zimbabwe; the range of African crops under study with molecular markers is very wide, however, from traditional commodities to tropical fruits. Of the different marker systems, RAPDs, microsatellites, RFLPs and AFLPs were commonly used. While the analysis does not distinguish between use of molecular markers for genetic improvement (marker-assisted selection) or for genetic characterization/conservation, it does indicate that, in crops at least, there is a wide range of marker-based research activities ongoing in developing countries, although some countries and regions seem to have far greater capacity than others.

The database also includes information on *in vitro* germplasm conservation and cryopreservation, indicating that there are only few activities in developing countries, although there may be substantial under-reporting. The report argues that low uptake of micropropagation techniques for *in vitro* germplasm conservation may also be due to the existence of established whole plant germplasm collections for species where *in vitro* conservation is appropriate, leading to a reluctance to provide funding for *in vitro* facilities. The balance between *in vitro* and whole plant collections may change as facilities for and capabilities in *in vitro* conservation increase and existing plant collections need rejuvenation (FAO, 2005).

Some of the specific questions on this issue that participants might wish to address in the e-mail conference might include:

- How important is this issue of capacity in different countries or regions?
- What are the current limiting factors to capacity building in developing countries? How can these limiting factors be overcome?
- What role should international organizations, such as FAO, the World Bank, or the Consultative Group on International Agricultural Research (CGIAR) research centres have in this area?
- What role should public-private partnerships play in this area?
- How important should regional capacity-building projects be with respect to national initiatives?

15.5.2 Economic issues

While obviously dependent on a range of factors such as the species or the size of the initiative, the use of biotechnology for characterization/conservation of genetic resources can be expensive.

For example, in a survey of livestock molecular marker biodiversity projects, respondents also provided information on costs of 41 projects (FAO, 2004b). They ranged from US$500 to 14 million, averaging US$ 130 000. For projects providing a breakdown of costs, it was estimated that, on average, 65 percent of the total project expenses were spent on genotyping, about 20 percent on sampling animals, 10 percent on the statistical analysis and 5 percent on project coordination. There was also considerable variation between projects, with genotyping costs ranging from 20-90 percent of the total costs in individual projects.

Cryopreservation technologies also can bear a considerable cost. FAO (1997, Annex 1-2) concludes that "the techniques required for successful cryopreservation of non-orthodox seeds are sophisticated and implementing a cryopreservation programme requires trained technical staff, advanced plant tissue culture facilities and increased transportation/handling costs to assure fresh materials. These constraints may present a barrier to effective technology transfer to many developing countries".

Some of the specific questions on this issue that participants might wish to address in the e-mail conference could include:

- In a situation where genetic resources for food and agriculture are being lost rapidly and where financial and human resources to support conservation activities are often limited, how much emphasis should be given to the application of biotechnology in this area?
- Which applications should be prioritized in this situation and in which sectors/species?

15.5.3 Prioritization issues

While in an ideal world, all available conservation options should be used, for financial reasons reality often dictates that tough choices have to be made. Although alternative tools or strategies can complement each other, priorities often have to be made. This is also valid for the biotechnologies and conservation strategies described in this chapter.

Molecular markers

Given the large numbers of animal breeds and crop varieties in danger of becoming extinct and limited resources, governments may need to identify the breeds and varieties they consider to be most important for conservation purposes in terms of financial support and storage in gene banks, *inter alia*. A range of factors may then be considered for the different populations, including their degree of endangerment, their economic value, their impact on the environment, their scientific value, their social and cultural importance and their genetic diversity (can be studied using DNA molecular markers). Characterizing populations for any one of these factors can be time-consuming and costly. Results from FAO (2004b) suggest that many studies, at least in livestock, are characterizing populations for more than one factor at a time. In 56 of the 86 livestock genetic diversity projects surveyed (i.e. about two-thirds), additional breed characteristics such as production performance, morphological traits, disease resistance, behavioural traits and cultural values were also recorded.

If information on these different factors is available, the question is then raised on how best to prioritize the different sources of information for conservation purposes. This is a valid question for each of the agricultural sectors where, for example, efforts have been made to develop methodologies for prioritizing between brown trout populations for conservation using factors such as the relative risk of extinction, the socio-economic and scientific value and the genetic/evolutionary and ecological legacy of the different fish populations (Primmer, 2005).

Some of the specific questions on this issue that participants might wish to address in the e-mail conference could include:

- How should genetic differences be weighted with respect to non-genetic differences (cultural importance, current market value, degree of endangerment, etc.) when wishing to prioritize between populations for conservation purposes?
- How should differences between populations in genetic diversity at neutral loci (measured using molecular markers) be weighted with respect to differences in genetic differences for traits that are or have been under selection (estimated using production/adaptation data)?
- Which kinds of molecular markers or statistical analysis methods are most appropriate for the different kinds of population of interest?

Cryopreservation, in vitro culture and reproductive technologies

As outlined earlier, *in situ* and *ex situ* conservation represent two major strategies. Both have their disadvantages (e.g. *in situ* populations are vulnerable to disease, drought and human interference, etc. while germplasm storage in *ex situ* collections may be very technology-intense and thus require relatively sophisticated equipment and skilled labour) and advantages (*in situ* populations can adapt to environmental changes while *ex situ* collections can be stored indefinitely). Economic merits of one over the other may depend on the species involved and the kind of *ex situ* collection, among other considerations.

Some of the specific questions on this issue that participants might wish to address in the e-mail conference might include:

- When is it more appropriate to support biotechnology as a tool for *ex situ* conservation in preference to supporting *in situ* conservation?
- What kind of biological material should be cryopreserved, considering that the goal may be to eventually regenerate a living population? For example, in animals, should the focus be on semen, embryo, oocytes, or a combination thereof?

The birth of the Scottish sheep Dolly in 1996, produced through somatic cell cloning using udder cells and nuclear transfer, was a major scientific breakthrough, and since then, individuals from a number of other species have been cloned. How important is this biotechnology for conservation of animal genetic resources?

15.6 REFERENCES

Bartley, D.M. 2005. Status of the world's fishery genetic resources. In *Proceedings of the International Workshop on the Role of Biotechnology for the Characterisation and Conservation of Crop, Forestry, Animal and Fishery Genetic Resources.* (available at www.fao.org/biotech/docs/bartley.pdf).

Cardellino, R.A. 2005. Status of the world's livestock genetic resources. Preparation of the first Report on the State of the World's Animal Genetic Resources. In *Proceedings of the International Workshop on the Role of Biotechnology for the Characterisation and Conservation of Crop, Forestry, Animal and Fishery Genetic Resources.* (available at www.fao.org/biotech/docs/cardellino.pdf).

de Vicente, M.C., Guzmán, F.A., Engels, J. & Ramanatha Rao, V. 2005. Genetic characterization and its use in decision-making for the conservation of crop germplasm. In *Proceedings of the International Workshop on the Role of Biotechnology for the Characterisation and Conservation of Crop, Forestry, Animal and Fishery Genetic Resources.* (available at www.fao.org/biotech/docs/vicente.pdf).

Diamond, J. 2002. Evolution, consequences and future of plant and animal domestication. *Nature,* 418, 700-707. (also available at www.nature.com/cgi-taf/DynaPage.taf?file=/nature/journal/v418/n6898/full/nature01019_fs.html&content_filetype=pdf).

ERFP. 2003. Guidelines for the constitution of national cryopreservation programmes for farm animals. *In* S.J. Hiemstra, ed. *The European regional focal point on animal genetic resources. Publication no. 1.* (also available at www.zum.lt/agroweb/Tekstai/Guidelinest.pdf).

FAO. 1997. *The state of the world's plant genetic resources for food and agriculture.* (also available at www.fao.org/ag/AGP/AGPS/PGRFA/pdf/swrfull.pdf).

FAO. 1998. *Secondary guidelines for development of national farm animal genetic resources management plans: management of small populations at risk.* Rome. (also available at dad.fao.org/en/refer/library/guidelin/sml-popn.pdf).

FAO. 1999. *The global strategy for the management of farm animal genetic resources: executive brief.* Rome. (also available at dad.fao.org/en/refer/library/idad/ex-brf.pdf).

FAO. 2000a. *How appropriate are currently available biotechnologies for the forestry sector in developing countries?* Background Document to Conference 2 of the FAO Biotechnology Forum (25 April to 30 June 2000). Rome. (available at www.fao.org/biotech/C2doc.htm).

FAO. 2000b. *World watch list for domestic animal diversity.* 3rd edition. Rome. (also available at dad.fao.org/en/refer/library/wwl/wwl3.pdf).

FAO. 2001. *Global forest resources assessment 2000.* Rome. (also available at www.fao.org/forestry/site/fra2000report/en).

FAO. 2002. *Gene flow from GM to non-GM populations in the crop, forestry, animal and fishery sectors.* Background Document to Conference 7 of the FAO Biotechnology Forum (31 May to 5 July 2002). Rome. (available at www.fao.org/biotech/C7doc.htm).

FAO. 2003. *Molecular marker assisted selection as a potential tool for genetic improvement of crops, forest trees, livestock and fish in developing countries.*

Background document to Conference 10 of the FAO Biotechnology Forum (17 November to 14 December 2003). Rome. (available at www.fao.org/biotech/C10doc.htm).

FAO. 2004a. *The state of world fisheries and aquaculture.* Rome. (also available at www.fao.org/DOCREP/007/y5600e/y5600e00.htm).

FAO. 2004b. *Measurement of domestic animal diversity - a review of recent diversity studies.* Information Document prepared for the Meeting of the Intergovernmental Technical Working Group on Animal Genetic Resources for Food and Agriculture, 31 March to 2 April 2004, Rome, Italy. (available at dad.fao.org/en/refer/library/reports2/itwg/CGRFA_WG_AnGR_3_04_Inf3.pdf).

FAO. 2005. *Status of research and application of crop biotechnologies in developing countries: preliminary assessment* by Z. Dhlamini, C. Spillane, J.P. Moss, J. Ruane, N. Urquia & A. Sonnino. Rome. (also available at ftp.fao.org/docrep/fao/008/y5800e/y5800e00.pdf).

Ferreira, M.E. 2005. Molecular analysis of genebanks for sustainable conservation and increased use of crop genetic resources. In *Proceedings of the International Workshop on the Role of Biotechnology for the Characterisation and Conservation of Crop, Forestry, Animal and Fishery Genetic Resources.* (available at www.fao.org/biotech/docs/ferreira.pdf).

Greer, D. & Harvey, B. 2004. *Blue genes: sharing and conserving the world's aquatic biodiversity.* Ottawa, Canada, International Development Research Centre. (also available at web.idrc.ca/openebooks/157-4/).

Hanotte, O. & Jianlin, H. 2005. Genetic characterisation of livestock populations and its use in conservation decision-making. In *Proceedings of the International Workshop on the Role of Biotechnology for the Characterisation and Conservation of Crop, Forestry, Animal and Fishery Genetic Resources.* (available at www.fao.org/biotech/docs/hanotte.pdf).

Hiemstra, S.J., van der Lende, T. & Woelders, H. 2005. The potential of cryopreservation and reproductive technologies for animal genetic resources conservation strategies. In *Proceedings of the International Workshop on the Role of Biotechnology for the Characterisation and Conservation of Crop, Forestry, Animal and Fishery Genetic Resources.* (available at www.fao.org/biotech/docs/hiemstra.pdf).

Lanteri, S. & Barcaccia, G. 2005. Molecular markers based analysis for crop germplasm preservation. In *Proceedings of the International Workshop on the Role of Biotechnology for the Characterisation and Conservation of Crop, Forestry, Animal and Fishery Genetic Resources.* (available at www.fao.org/biotech/docs/lanteri.pdf).

Lenstra, J.A. & Econogene Consortium. 2005. Evolutionary and demographic history of sheep and goats suggested by nuclear, mtDNA and Y-chromosome markers. In *Proceedings of the International Workshop on the Role of Biotechnology for the Characterisation and Conservation of Crop, Forestry, Animal and Fishery Genetic Resources.* (available at www.fao.org/biotech/docs/lenstra.pdf).

Panis, B. & Lambardi, M. 2005. Status of cryopreservation technologies in plants (crops and forest trees). In *Proceedings of the International Workshop on the Role of*

Biotechnology for the Characterisation and Conservation of Crop, Forestry, Animal and Fishery Genetic Resources. (available at www.fao.org/biotech/docs/panis.pdf).

Papa, R. 2005. Gene flow and introgression between domesticated crops and their wild relatives. In *Proceedings of the International Workshop on the Role of Biotechnology for the Characterisation and Conservation of Crop, Forestry, Animal and Fishery Genetic Resources.* (available at www.fao.org/biotech/docs/papa.pdf).

Primmer, C. 2005. Genetic characterisation of populations and its use in conservation decision-making in fish. In *Proceedings of the International Workshop on the Role of Biotechnology for the Characterisation and Conservation of Crop, Forestry, Animal and Fishery Genetic Resources.* (available at www.fao.org/biotech/docs/primmer.pdf).

Rao, N.K. 2004. Plant genetic resources: advancing conservation and use through biotechnology. *African Journal of Biotechnology*, 3: 136-145. (also available at www.academicjournals.org/AJB/PDF/Pdf2004/Feb/Rao.pdf).

Ruane, J. 2000. A framework for prioritizing domestic animal breeds for conservation purposes at the national level: a Norwegian case study. *Conservation Biology*, 14: 1385-1393. (also available at www.nordgen.org/ngh/download/bokartikkel-ruane.doc).

Sigaud, P. 2005. Efforts towards assessing the global status of forest genetic resources. In *Proceedings of the International Workshop on the Role of Biotechnology for the Characterisation and Conservation of Crop, Forestry, Animal and Fishery Genetic Resources.* (available at www.fao.org/biotech/docs/sigaud.pdf).

Simianer, H. 2005. Use of molecular markers and other information for sampling germplasm to create an animal genebank. In *Proceedings of the International Workshop on the Role of Biotechnology for the Characterisation and Conservation of Crop, Forestry, Animal and Fishery Genetic Resources.* (available at www.fao.org/biotech/docs/simianer.pdf).

16. Summary of discussion from the e-mail conference on the role of biotechnology for the characterization and conservation of crop, forest, animal and fishery genetic resources in developing countries

John Ruane and Jonathan Robinson

16.1 SUMMARY

Characterization and conservation of genetic resources of crops, forest trees, livestock and aquatic species are important for all countries, but particularly for developing countries whose economies depend heavily on these sectors, and where genetic resources are often threatened. A number of biotechnology tools are available that can help in characterization and conservation of such genetic resources, ranging from relatively cheap and uncomplicated technologies to sophisticated, resource-demanding ones. In each of the crop, forestry, animal and fishery sectors, albeit to different degrees, biotechnology tools are currently being applied in developing countries for these purposes and numerous examples of the wide range of applications were provided during this FAO e-mail conference. Of the different biotechnologies, most discussions were about molecular markers, in particular their use for characterization of genetic resources, where issues such as the advantages or disadvantages of different marker systems and the proposal to develop a universal molecular marker database were debated. In situations involving potential use of marker and non-marker information, such as development of a core collection of plant gene bank accessions or prioritization of animal breeds for conservation purposes, there was general consensus that decisions should not be based on marker information alone and that other factors, such as morphology and agronomic performance, should also be considered. The merits of several *in vitro* techniques, including tissue culture, cryopreservation and DNA storage, were considered with a view to conservation of genetic resources, where e.g. DNA banks for plants were seen as potentially complementing but not replacing seed banks, at least in the near future. The ability to apply these biotechnologies in developing countries is currently limited by the lack of sufficient funds, human capacity and adequate infrastructure. The importance of human resource capacity building was highlighted. There was a general call for greater collaboration among researchers and practitioners, particularly at the regional level, to reduce costs and pool limited resources, and between developed and developing country institutions. A role was seen for international organizations, including FAO, and the centres of the Consultative Group on International Agricultural Research (CGIAR), in coordinating these collaborative efforts and in supporting these capacity-building activities.

16.2 INTRODUCTION

The theme of this e-mail conference, which was hosted by the FAO Biotechnology Forum between 6 June and 3 July 2005, was the role of biotechnology for the characterization and conservation of crop, forest, animal and fishery genetic resources in developing countries. This chapter provides a summary of the principal issues and opinions received from contributors to the conference. Specific messages are referenced in the document using participants' surnames and the message number. All messages can be read at www.fao.org/biotech/logs/c13logs.htm. About 650 people subscribed to the conference and 127 e-mail messages were posted from people living in 38 different countries; over 60 percent of messages were from developing countries. Most of the messages came from people working in research organizations, including CGIAR centres, and universities.

Most participants directed their messages to issues in one of the crop, forestry, animal or fishery sectors, with greatest attention given to crops and animals. Some also addressed cross-sectoral issues such as resource availability and constraints and international collaboration. In Section 16.3 of this chapter, the main issues discussed during the conference are summarized. Section 16.4 provides information on participation and Section 16.5 represents a list of names and countries of those who sent referenced messages.

16.3 MAIN THEMES DISCUSSED

16.3.1 Current applications and potential of biotechnology in the different sectors

Most of the messages addressed the current situation and potential of using biotechnology tools for the characterization and conservation of genetic resources in one of the individual sectors (crop, forestry, animal or fishery). Although each sector has its specificities, some of the discussions were also very relevant for other sectors.

Crop genetic resources

The potential importance of biotechnology for locally important crop genetic resources was raised right at the beginning of the conference. Nkhoma (1), describing activities at the Southern African Development Community Plant

Genetic Resources Centre (SPGRC) in Zambia, noted that while their programme had succeeded fairly well with common cereals and legumes, they were in some cases unable to work on indigenous crops - for example when difficult to cultivate - which were locally useful and on which nobody else was working. He said that biotechnology facilities would help them to tackle these problems. He also reported that the SPGRC collections were only characterized using agronomic and morphological data, although molecular markers would be helpful to allow removal of duplicate accessions. The usefulness of applying biotechnology to genetic diversity studies of less common crops was reported by Infante (8), who described their results in Venezuela where they found genetic variability in clonally reproduced henequen and cocuy following analysis with molecular markers. This had implications for conservation because clonally reproduced species were assumed to be of uniform genetic constitution. Morphological data were also gathered and analysed which supported the molecular data.

Kisha (6) noted that molecular markers are now an accepted and widely used tool for measuring genetic diversity, where "molecular marker technology can be used to characterize the extent of diversity within a collection and for the development of collection management strategies, which may include establishment of core collections, identification of redundancies or contamination, guidance for future collection efforts, and identification of gaps of ancestral crop relatives. Additionally, analysis of world-wide genetic diversity can identify areas suited for the establishment of *in-situ* conservation sites."

Ndjiondjop (41) reported that recent studies with molecular markers based on repetitive DNA sequences had provided useful information on genetic diversity present in rice, *Oryza glaberrima*. Previous studies using isozyme and restriction fragment length polymorphism (RFLP) markers had failed to identify as many polymorphisms, although considerable variation for many morphological and agronomic traits had been recorded. The microsatellite marker information will be used to develop the core collection of *O. glaberrima* accessions in the Africa Rice Center (WARDA) gene bank. Also regarding core collections, Huaman (38) said that "molecular markers are the most valuable source to get data on genetic diversity of a given crop or plant species", but added that when selecting a core collection, other factors, such as eco-geographical data, disease and pest reaction and morphological diversity, had to be taken into account. Ghamkhar (28) suggested that molecular markers should be "definitely employed as the best technique for screening of genetic diversity but they must also be double checked by morphological data to make certain there is no major loss in our breeding and/or core collection development programs."

Molecular marker data were considered to be only one aspect of characterization (Huaman, 38), but one that would be of greater use were a universal molecular marker database to exist (Kisha, 6). He noted that in the current situation, few, if any, plant genetic diversity studies can be directly compared or compiled; that marker data can be lost or forgotten after publication;

and that studies of genetic diversity are usually limited to a few accessions or accessions from a limited area of interest. The development of such a database was supported by several participants, including Vijay (18), who urged, however, that consideration should be given to selection of a set of universally reliable and reproducible markers; to consensus on the outcome from different markers; and to standardization of methodologies, including the mode of analysis. In this context, both de Vicente (26) and Ford-Lloyd (30) mentioned an initiative by the International Plant Genetic Resources Institute (IPGRI) to define community standards for documenting information about genetic markers so that researchers can generate and exchange genetic marker data that are standardized and replicable. Barker (24) noted that a large number of molecular markers were already available in the public domain, but to assemble a universal database would require considerable effort and it was hard to see any single country willing to invest in such a project. He did envisage, however, that it would be easier to establish databases on a single species basis. Kisha (51, 103) expanded on his original proposal, describing how the database would need to be curated and why a core set of primers would be useful to make studies comparable. Sales (19) noted that some microsatellite databases already exist and Ghamkhar (28) mentioned some Web-based databases, mostly for cereals. R. Jones (54) hoped that a global marker database might also include data for key fish and crustacean species.

Kisha (51) also commented on the relative usefulness of different types of molecular markers, arguing that amplification fragment length polymorphism (AFLP) markers can cover a large area of the genome with less cost than simple sequence repeat (SSR) markers, also known as microsatellites. Ghamkhar (28, 80) proposed that inter-simple sequence repeat (ISSR) techniques were also cheap and efficient, and Varshney (43) supported using single nucleotide polymorphism (SNP) markers for characterization, while acknowledging their high cost and the relative paucity of species for which markers currently existed. In a similar vein, Ghamkhar (53) suggested that SNPs might be useful in well-studied crops such as wheat and maize but not for less common crops because of the sequencing work required to get the markers. Warburton (42) described experiences in using SSR markers to study molecular diversity in maize among several laboratories and highlighted problems of comparison and reproducibility among laboratories even when using the same protocols and platforms for genotyping. She said that the possibility of combining datasets was "virtually non-existent" if laboratories used different techniques and that there were problems of repeatability with some other kinds of markers as well. Varshney (43) had had similar experiences and suggested that expressed sequence tag (EST)-derived SSR markers showed higher reproducibility than genomic SSR markers. Krishna (88), supported by Buso (93), felt that all molecular marker data were nevertheless useful, and even though random amplified polymorphic DNA (RAPD) and AFLP markers might have some problems with reproducibility (highlighted by Vijay [79]), they also had advantages (highlighted by Muchugi [77]).

Dulieu (95) also compared different markers systems with respect to the reproducibility of their results, noting that some markers, such as microsatellites, were more reliable than others but required more preliminary research, and suggested that biotechnology companies should be encouraged to produce kits for the most important species, following the example of human DNA fingerprinting kits which are used universally. De Vicente (26) noted that as part of the CGIAR Generation Challenge Program, microsatellite kits were being put together which they hoped to have available in the near future. Dulieu (95) also pointed out that many of the molecular markers revealed differences between populations that were not highly correlated with performance or phenotypic characters. Gupta (44) suggested that if molecular markers were to be used for genetic diversity analysis, they should be functional markers rather than random genomic markers. He (44, 87) also reported that they got different results from genetic distance analyses of bread wheat when different kinds of markers (SSR, AFLP or selective amplification of microsatellite polymorphic loci [SAMPL]) were used, suggesting that the best estimate of diversity might be obtained from data on large numbers of morphological traits. Kisha (88) wrote that it would be useful to compare at least two different marker systems for agreement in the resulting relationships. Ghamkhar (118) agreed that there can be inconsistencies between results obtained with different sets of molecular and morphological data, concluding, "more molecular techniques/data, more resolution or better results."

The merits of tissue culture as a means of genetic resource conservation were discussed by several participants (e.g. Muchugi, 68). Lin (2) suggested that tissue culture and other forms of micropropagation were useful tools for the conservation and multiplication of plant species, noting that low-cost options were also available. Cummins (9), however, felt that because of somaclonal variation (i.e. mutations that occur spontaneously in tissue culture), tissue culture was not a good way of conserving local genetic material. Muralidharan (22) agreed and favoured use of slow-growing shoot culture. He suggested also that molecular markers could be used to study the extent of somaclonal variation in slow-growing or cryopreserved cultures. Lin (21) reported that improved protocols for *in vitro* conservation, developed for a range of species, could overcome the potential problems of somaclonal variation and that *in vitro* conservation was useful, particularly for plants that do not produce seeds or that produce seeds of limited viability. Ford-Lloyd (30) pointed out that *in vitro* conservation was being used to support genetic conservation, despite genetic instability, as it represented a better option than the currently available alternatives.

There were several responses to a question from Muralidharan (22) regarding the potential of DNA as a means of long-term conservation of genetic material. Wang (32) pointed out that germplasm conservation as pure DNA was already a reality in some countries, which was supplemented by concrete examples in later messages (Ghamkhar, 38, 48; Widjaja, 40; Vijay, 47). De Vicente (46) reported results from a 2004 worldwide survey on plant genetic resources DNA banking activities showing that 20 percent of the 243 institutions that replied to the questionnaire kept DNA as a genetic resource. Although noting that both DNA banks and seed banks have advantages and disadvantages, Ghamkhar (34) suggested that seed banks were currently preferable to DNA banks since, for example, contamination was more immediately apparent and morphological screening and maintenance were easier. Vijay (47) agreed with Ghamkhar (34) and described the major limitations he saw to the use of DNA banks, suggesting they could complement, but not replace, seed banks, at least in the near future.

Dulieu (96) preferred phage genomic libraries over DNA banks for genetic resource conservation because they are easier to prepare and maintain. However, he warned against using more technology to counter effects of misuses of technology (mainly the destruction of traditional agricultural systems), a point echoed in a different context by Magalhães (72), supported by Kante (105) and Adediran (112), who argued that the use of biotechnology to conserve or characterize biodiversity could not be considered a solution but only a palliative when biodiversity in developing countries was being destroyed by an economic model based on the economic exploitation of developing countries and their natural resources.

Nassar (4) suggested that apomixis (i.e. where seeds are produced through asexual processes so that the genetic make-up of the seeds is identical to that of the mother plant) could contribute to conservation of certain crop genetic resources and reported that they had produced apomictic cassava clones in Brazil, confirmed using molecular markers. Vijay (11) agreed and felt that identification and transfer of apomixis to other cultivated species would have important consequences for conservation and hybrid seed production.

During the conference, Gupta (23, 87), supported by Vijay (73), also mentioned the new possibilities of using DNA barcoding, allowing different plant species to be identified and discriminated. Ghamkhar (118) felt it could help taxonomists to classify, re-classify or identify taxa or new species, when used together with traditional taxonomy methodologies based on morphological data. Gupta (120) also pointed out the potential of DNA microarrays for the study of genetic diversity.

Forest genetic resources

There were relatively few contributions specifically addressing the use of biotechnology in characterization and conservation of forest genetic resources. Oluawsegun (101) reported on the loss of forest species in Nigeria due to factors such as poverty and low education levels among the rural people, concluding "for any meaningful and lasting conservation programme to be effectively carried out there must be a conscious effort in involving the local people in maintaining and managing their environment since the needs of these dwellers must be respected." Muralidharan (67) bemoaned that there were insufficient effort and funding put into conservation of forest genetic resources compared to the crop and livestock sectors and wondered whether tropical forest genetic resources could be successfully conserved in DNA banks as they "are in danger of mass erosion due to degradation of the habitat." He emphasized, however, that wherever feasible, the more conventional conservation methods should be used. Ghamkhar (76) shared his concerns and supported the use of *ex situ* conservation methods (seed banks, storing tissues, or DNA banks) as there did not seem to be other options.

Muchugi (68) was also concerned about conservation of indigenous forest genetic resources, which are threatened by factors such as increasing population sizes (requiring land for settlement and farming). Although biotechnology could be of value, through e.g. use of molecular markers to investigate gene flow or to assist in establishment of *ex situ* conservation programmes, she noted that little work had been carried out on tropical tree species compared to temperate species. She (68, 77) described the advantages, in terms of relative costs and speed, of using RAPD markers and isozymes for molecular characterization of tropical tree species, but was saddened that renowned molecular genetics journals refused to publish results using these simpler techniques, concluding: "this is placing scientists in the developing world with simple labs in a tricky position; we would love to employ the modern state of art sequencers but financial limitations will not allow it. What is the way out then considering the need to study these taxa before we lose them on the earth's surface." The importance of molecular markers for management of endangered natural forest tree species was also underlined by Dulieu (96).

Animal genetic resources

Aziz (7), supported by Silva (20), noted that livestock production in developing countries is characterized by several major constraints, such as the absence of national recording systems, paucity of breeding programmes, small herd sizes, lack of awareness of the importance of animal genetic resources, national policies to replace local breeds with exotic ones, scarce resources and weak infrastructure. He (82) proposed that in developing countries a systematic approach be taken to the documentation, evaluation, conservation and utilization of their animal genetic resources. Babar (107, 115) emphasized the importance of conserving local breeds in developing countries, especially in countries such as Pakistan that are home to several important breeds. He (107) encouraged use of cryopreservation technologies, including embryo and semen storage, and establishment of DNA banks (as they had done in Lahore). Sales (84) said that developing countries had been "swamped" with new livestock breeds and little had been done to conserve their native breeds. All these issues represented obstacles to the application of biotechnology, as communicated by Silva (20) and Tantia (65).

Hassan (61) described the limitations to livestock production in sub-Saharan Africa, which were similar to those previously outlined by Aziz (7). He suggested that although molecular genetics might have a great role to play in revolutionizing livestock production in developing countries, "the stage is not yet set" for them to do so and that developing countries should continue the ongoing phenotypic characterization. Maddul (89) mentioned their work on phenotypic characterization and conservation of pigs and chickens in the northern Luzon highlands of the Philippines, where molecular characterization had not been carried out due to the lack of a laboratory. Tan (12) described results from molecular characterization studies of buffaloes in Southeast Asia carried out in the 1990s. Silva (20) reported that in Sri Lanka, applications of biotechnology are limited, involving artificial insemination (AI) and, in isolated cases, genetic characterization of local animals. Osakwe (104) suggested that although some progress had been reported in plants, there "is no meaningful characterisation and conservation of genetic resources using biotechnology in animals in most developing countries."

Köhler-Rollefson (31) felt that although scientifically interesting, the relevance of molecular characterization of livestock breeds for livestock keepers and for poverty alleviation had yet to be proven. She also asked whether, when studying livestock domestication and dispersal, molecular data were superior to data from archaeological and ethno-historical investigations, or if they merely confirmed what was already known. Lenstra (35) said that there were several examples where DNA studies had provided additional insights into these issues. For Hanotte (36), molecular characterization had provided an additional source of information that, together with data from archaeology, linguistics and indigenous knowledge, was making it possible to "complete a puzzle about origin and history of agriculture." P. Jones (59) reported that recent molecular work had substantiated the long-held belief that most of today's donkeys have their origins in northern Africa and that those in southern Africa have different origins, concluding "such a history has implications for management practices and technology transfer, and thus indirectly on poverty alleviation."

Köhler-Rollefson (31) also asked whether genetic uniqueness should be the key criterion for deciding which breeds to prioritize for conservation purposes. Lenstra (35) argued that such a decision should not be made on the basis of marker data alone and that other factors, such as the breeds' relevance to regional tradition, were also important. Tantia (65) thought that molecular characterization with microsatellites was an excellent tool in this context. Toro (71) argued that the important parameters to consider were phenotypic ones, adaptation to specific environments, possession of economically important traits and cultural/historical value, etc., and that "to use molecular markers is not harmful as long as we recognize its subordinate relevance in this context." Hanotte (36) suggested that molecular criteria should be considered very important when selecting animals for *ex situ* conservation, but caution should be advised for *in situ* conservation at the farmer community level. Nimbkar (45) argued that areas directly related to poverty alleviation, such as characterizing the performance of indigenous breeds (leading to genetic improvement programmes based on simple principles and methods of animal breeding), were being neglected in developing countries such as India in favour of biotechnologies, such as molecular characterization or embryo transfer, blaming this trend on the "glamour" of these biotechnologies. This view was supported by Steane (55), who suggested that one of the problems is that in most countries the funding bodies emphasize research rather than application, concluding "certainly biotech has an important role but it will never be the sole criterion for decisions about conservation and characterisation of breeds." On the general issue of weighting genetic versus non-genetic differences, Rakotonjanahary (92) felt the former should be ranked at a lower level because "socio-economic traits such as culture, market value, degree of endangerment are more related to humankind, which is to be considered the top priority."

Galal (60, 66) suggested that it would be useful for conservation purposes to use marker data to verify if phenotypically similar breeds were actually the same or not, although Toro (63, 71) did not see how this could be verified since molecular markers refer to neutral or non-coding genetic variation and genetic distances ignore within-breed variation. Lenstra (70), based on results from two recent European Union research projects, suggested that phenotypic distinctness and molecular diversity could both represent valid reasons for conservation, but they were often negatively correlated. Toro (99) argued that genetic distances based on molecular markers were only partially informative, and that there had been too much emphasis on molecular markers, with very few papers published on phylogeny based on phenotypic information compared to "hundreds" on genetic distances using molecular markers. Ghamkhar (100), in response, listed several advantages of molecular markers over morphological data for conservation purposes, including their relative abundance, stability under changing environmental conditions, consistency among laboratories, reduced time requirement and ease of data analysis. Chagunda (113) saw a need for decision support tools to assist in animal genetic resource conservation strategies, where information from different sources (molecular, phenotypic, genotypic, production system, indigenous knowledge, socio-economics, etc.) could be combined in a meaningful way to support decisions for conservation strategies.

Fishery genetic resources

R. Jones (14, 54) felt that "fisheries resources are undervalued and underrepresented in discussions on the conservation and sustainable use of germplasm" and that "the application of molecular tools in the relatively unknown world of fisheries and aquatic biodiversity will continue to play roles in determining stock structures of multi-species fisheries (meta-population dynamics) and important taxonomic classification work". Following the message of Oluawsegun (101), describing the pressures on natural habitats in developing countries, R. Jones (102) urged that those working on conservation of genetic resources in developing countries should think seriously about how to prioritize their efforts. He (64) noted that because of the high costs involved, most of the molecular work in fisheries was done on commercial or potentially commercial species and wondered how these results might be applied to alleviation of hunger and poverty.

Tan (12) wrote that microsatellites were being used to study three commercial aquatic species in Malaysia - the giant freshwater prawn, the green-lipped mussel and the Asian river catfish. Chinsembu (37) described recent research on the mitochondrial DNA analysis of cichlid fish from five southern African rivers. Results of this work had helped to explain details of the evolution and radiation of the species. Bhassu (62) was enthusiastic about the impacts of biotechnology in stock improvement and characterization of fish, citing recent work on characterization of red tilapia stocks with microsatellites; on the phylogeny of freshwater prawns (where mitochondrial DNA analysis confirmed previous findings from morphological and protein data); and on the use of AFLP markers for sex determination in freshwater prawns, noting, however, that little funding was available for such research. Ablan (75) felt that molecular analysis was not as meaningful for cultivated populations as for wild populations, for which they can define spatial distribution of stocks and provide guidance on where to obtain brood stock for restocking programmes. She noted that "we've had cases where molecular genetics provide answers, others where it simply validates observation of the phenotype, local knowledge, value judgements, or even gut feel. And yes, sometimes its usefulness can be forced."

Tan (69) reported that protein and molecular marker analysis had indicated low levels of genetic diversity in many cultured stocks of prawn and sea bass, which were associated with abnormalities and reduced fertility. He argued that such results had important economic implications, justifying investments in genetic marker studies, and that stock deterioration could therefore be halted based on this knowledge. Toro (71), arguing that the phenomenon of inbreeding depression was already well known, was not convinced that molecular markers would provide useful additional information here. Tan (81), however, noted that aquaculturists in Malaysia usually did not maintain breeding records, paid little heed to inbreeding and so were often surprised when molecular studies revealed that their problem stocks were inbred. Only after seeing the molecular typing results would they begin outcrossing. Kalamujic (110) reported that in Bosnia and Herzegovina, molecular marker characterization of salmonid species had led to a new freshwater fisheries law, including a provision on obligatory genetic control of material for stocking. R. Jones (14) described how DNA fingerprinting of sturgeon had helped to provide information on stock make-up, potential loss of genetic diversity in Russian brood stocks, the caviar trade and use of *ex situ* conservation.

16.3.2 Priorities and resource constraints in developing countries

Some participants discussed the priority that developing countries give or should give to applying biotechnology for the characterization and conservation of their genetic resources. For example, Qureshi (3) wrote that conservation of animal genetic resources was neglected in developing countries as the state had failed to support it and it was also not a priority for the farmers. He saw, however, that in this situation, universities could play an important role in applying biotechnology in this area. Huque (52) suggested that policy-makers in developing countries did not support animal biotechnology as they preferred to invest in areas promising short-term rather than long-term benefits. Rakotonjanahary (92) said the weak capacity to use biotechnology for characterization/conservation in developing countries was understandable given that, in general, national policy prioritized poverty alleviation, food self-sufficiency and increased agricultural productivity over research activities. Komwihangilo (57) argued that farmers in, for example, sub-Saharan Africa were not concerned about characterization but about their livelihoods and their animals being able to produce. Therefore, molecular characterization and other new technologies will only be of value to the farmers "if they deliver them from the present trials of life and assure them sustainable futures."

As most of the world's poor live in the rural areas of developing countries, Djoulde (16) was skeptical about the merits of using advanced technologies, such as molecular markers or cryopreservation and reproductive technologies, in these areas. De Vicente (27) wondered whether work should therefore cease on solving such problems if the rural people did not care about the solutions and whether effort should be concentrated on ensuring that the research findings reach the rural people. Djoulde (29) supported this latter point and emphasized that the complicated technologies needed to be adapted for easier application in developing countries. Ghamkhar (33) suggested that farmers in rural areas were not expected to employ the new methodologies, but should benefit from the better adapted germplasm resulting from their application. Considering the instability and threats that many gene banks faced in developing countries (due to human or environmental factors, such as war, hurricanes or famine), Murphy (124) questioned the merits of progressing to advanced methods such as molecular markers or tissue culture, concluding, "in places where seed banks are being destroyed by looters who are just after the plastic bottles, or where stocks are dying from want of electricity to refrigerate them, we need to question our priorities".

Many participants (e.g. Nkhoma, 1; Krishna, 74) commented on the lack of financial, human and infrastructural resources in developing countries for applications of biotechnology for the characterisation and conservation of genetic resources. As a typical example, Huaman (38) highlighted the difficulties facing researchers in developing countries wishing to use molecular markers, due to lack of laboratory equipment or materials. Aziz (7) and Galal (60) noted that the high costs of biotechnology, along with training, equipment and infrastructure, directed at livestock, represented a constraint. Komwihangilo (57) pointed out that the problem was not biotechnology-specific as there were, in general, low levels of funding for agricultural research. Oluawsegun (101) said economic problems meant that his government was "unable to allocate enough resources for conservation, for research and monitoring of conservation programmes, for the creation of gene banks and education of the public concerning the importance of preserving the biosphere." Urriola (50) reported that despite insufficient funding, human capacity, equipment and infrastructure, researchers in Panama were nevertheless trying to take advantage of the biotechnology resources available. Murphy (124) indicated that there were also serious resource shortages in the flagship centres of the CGIAR.

Despite the constraints, some participants urged that developing countries should not be left behind. For example, Vijay (73) considered that, despite the large amount of resources needed, it was important to keep up with the advancing technology: "Developing countries cannot stay aside from the mainstream knowledge as most of the diversity and its end users belong to them." Similarly, Kapoor-Vijay (117) urged that developing countries should not become "technologically excluded" and that "information, knowledge, and expertise associated with biotechnology which is relevant and needed to conserve unique plant, animal and microbial species thriving in diverse and especially extreme environments should be strengthened." For Edema (123), although developing countries did not, on their own, have enough resources to advance in biotechnology, they could not afford to be left behind either. R. Jones (15), on the other hand, warned of the dangers of being seduced by technology in a developing country context, "where the capacities to understand, absorb and if necessary fix and upgrade may be limited or non-existent." He emphasized, however, that no country should be deprived of the knowledge of the technologies available and that it was their choice if or how to apply them.

Not all biotechnologies are, however, equally resource-demanding. Lin (2) commented on the high costs of establishing and operating tissue culture and micropropagation facilities in developing countries, but pointed out that low-cost options had been developed. Thro (106) described one such low-cost initiative for propagating Andean root crops in rural areas. R. Jones (15) commended work being done on development of low-cost, portable cryopreservation technology for fish gametes. Kisha (98), commenting on the application of molecular markers, suggested that inexpensive, high throughput technology should be a primary consideration. Muchugi (68, 77) proposed the use of cheaper marker systems, such as RAPDs, in preliminary studies to form the basis of conservation strategies. Vijay (79), however, disagreed because of problems of reproducibility of RAPDs, concluding that although funding was the main problem for scientists in developing countries, the "use of outdated technology because of its low cost is not the answer." He (49) emphasized the importance of adoption, moving from "lab to land", to make technology useful for the common good, and that further research could allow the technologies developed to be made user-friendly. Similarly, Prana (58) urged a down-to-earth approach, focusing on appropriate technology adjusted for the real-life resources, local needs and socio-economic factors.

The importance of human capacity was highlighted with, for example, Prana (56) urging that human resource development be placed highest on the priority list. Uzochukwu (83), supported by Krishna (86) and Osakwe (104), argued that the large funding bodies "are interested in providing financial support for biotechnology research but not for training and updating local scientists in the developing countries. Krishna (74), supported by Uzochukwu (83), called for a massive capacity-building effort at the national and international level. Caesar (126) outlined the key features of a potential global biotechnology capacity-building project, based on regional and subregional groupings of developing countries and including a comprehensive scholarship/fellowship programme for developing countries. Sales (84) identified a need for developing country scientists to be regularly updated on the current trends and innovations relevant in biotechnology. She cited the Asian Maize Biotechnology Network, established to strengthen the biotechnology capacity of national maize research programmes in Asia, as a good model that, for example, livestock scientists might follow. Ghamkhar (85) also pointed out that the CGIAR Generation Challenge Program, previously described by de Vicente (26), provided relevant training programmes. He argued, however, that developing countries could not expect international organizations to cover all their training needs and that the countries should prepare a strategic plan and national training programme to transfer the knowledge accumulated by their already-trained senior scientists to the national research centres and universities.

16.3.3 Cooperative approaches

A recurring theme throughout the conference was that biotechnology research and application of results to characterization and conservation of germplasm can benefit from collaborative efforts, particularly at the regional level (e.g. Silva, 20; Ghamkhar, 78). In livestock, Galal (60), noting that the costs of equipment and materials for molecular characterization were too high for most local institutions, argued that "some regional coordination, possibly with international input, is required to carry out such work." In a similar vein, Muchugi (68), arguing that molecular characterization of tree species was best approached from an ecological/geographical perspective as their distributions cut across political boundaries, called for "greater collaboration among scientists within the regions in exchange of plant materials and knowledge gathered." The importance of collaboration between developed and developing country institutions was also highlighted (e.g. Prana, 56; Babar, 107, 115), with Vijay (121) arguing that "with the advancement of technology (like use of molecular markers for conservation) there should be a proper collaboration between these two parts of the world." Rakotonjanahary (92) thought that public-private partnerships in developing countries could play an important role, but "only on the condition that characterization and conservation have an evident economic impact, which is not the case for the moment."

Suggestions for increased cooperation were generally made in the interests of pooling scarce resources and reducing costs (e.g. Galal, 60; Muchugi, 94; Chinsembu, 108). De Vicente (25), arguing that it was not realistic for all countries to have facilities to carry out their own work, proposed that institutions consider the possibility of having a hub centre in their region where they could either send their samples for analysis or go there to do the work. Muchugi (94) cited the Biosciences Eastern and Central Africa (BECA) facilities in Nairobi as a good example to show how pooling of resources could help in the advancement of biotechnology. A universal molecular marker database was suggested by Kisha (6), based on collaboration among germplasm conservation centres and other interested parties. This idea was supported by several participants (e.g. Sales, 19), with Barker (24) emphasizing the requirement for "international collaboration and for the data to be maintained in the public domain." The role of international organizations, such as FAO, and the CGIAR centres, in coordinating these collaborative efforts, providing funds and contributing to capacity building was emphasized in several messages (e.g. Ghamkhar, 53, 78; Muchugi, 68; Vijay, 73, 79; Rakotonjanahary, 92; Babar, 107; Caesar, 126).

16.4 PARTICIPATION

The conference ran for four weeks, from 6 June to 3 July 2005. There were 645 subscribers to the conference, of whom 64 (10 percent) submitted at least one message. There were 127 messages in total, of which 61 percent came from people living in developing countries. Contributions to the conference came from all major regions of the world, with 28 percent of messages from Asia, 20 percent from Africa, 17 percent from Europe, 13 percent from Latin America and the Caribbean, 13 percent from North America and 10 percent from Oceania. Contributors represented 38 countries, the greatest numbers of messages coming from India, Australia, Canada, Brazil, United States, France, Kenya, Malaysia and Nigeria respectively. Most of the messages came from people working in research organizations (45 percent), including centres of the CGIAR, and universities (43 percent). The remainder were from independent consultants or from people working for an inter-governmental institute, non-governmental organization, national development agency or private company.

16.5 NAME AND COUNTRY OF PARTICIPANTS WITH REFERENCED MESSAGES

Ablan, Menchie. Malaysia
Adediran, Samuel Adeniyi. Gambia
Aziz, Mahmoud Abdel. Egypt
Babar, Masroor Ellahi. Canada
Barker, Guy. United Kingdom
Bhassu, Subha. Malaysia
Buso, Glaucia Salles Cortopassi. Brazil
Caesar, John. Guyana
Chagunda Mizeck. Denmark
Chinsembu, Kazhila Croffat. Namibia

Cummins, Joe. Canada
de Vicente, Carmen. Colombia
Djoulde, Darman Roger. Cameroon
Dulieu, Hubert. France
Edema, Olayinka. Nigeria
Ford-Lloyd, Brian. United Kingdom
Galal, Salah. Egypt
Ghamkhar, Kioumars. Australia
Gupta, P.K. India
Hanotte, Olivier. Kenya
Hassan, W. Akin. Nigeria
Huaman, Zosimo. Peru
Huque, Quazi M. Emdadul. Bangladesh
Infante, Diogenes. Venezuela
Jones, Peta. South Africa
Jones, Ron. Canada
Kalamujic, Belma. Bosnia and Herzegovina
Kante, Bocar. Italy
Kapoor-Vijay, Promila. Switzerland
Kisha, Theodore. United States of America
Komwihangilo, Daniel. United Republic of Tanzania
Köhler-Rollefson, Ilse. Germany
Krishna, Janaki. India
Lenstra, Hans. Netherlands
Lin, Edo. France
Maddul, Sonwright. Philippines
Magalhães, Vladimir. Brazil
Muchugi, Alice. Kenya
Muralidharan, E.M. India
Murphy, Denis. United Kingdom
Nassar, Nagib. Brazil
Ndjiondjop, Marie Noelle. Benin
Nimbkar, Chanda. Australia
Nkhoma, Charles. Zambia
Oluawsegun, Adegoke Adedayo. Nigeria
Osakwe, Isaac. Nigeria
Prana, Made Sri. Indonesia
Qureshi, Muhammad Subhan. Pakistan
Rakotonjanahary, Xavier. Madagascar
Sales, Emma. Philippines
Silva, Pradeepa. Sri Lanka
Steane, David. Thailand
Tan, S.G. Malaysia

Tantia, M.S. India
Thro, Ann Marie. United States of America
Toro, Miguel. Spain
Urriola, Jazmina. Panama
Uzochukwu, Sylvia. Nigeria
Varshney, Rajeev. Germany
Vijay, D. India
Wang, Richard. United States of America
Warburton, Marilyn. Mexico
Widjaja, Elizabeth. Indonesia